REPEATED MEASURES DESIGN FOR EMPIRICAL RESEARCHERS

REPEATED MEASURES DESIGN FOR EMPIRICAL RESEARCHERS

J. P. VERMA

Centre for Advanced Studies
Lakshmibai National Institute of Physical Education
Gwalior, India

Library of Congress Cataloging-in-Publication Data:

Verma, J. P.
    Repeated measures design for empirical researchers / J.P. Verma.
        pages cm
    Includes index.
    ISBN 978-1-119-05271-5 (cloth)
    1. Science–Methodology. 2. Experimental design. I. Title.
    Q175.V4235 2015
    001.4'2–dc23

                                                                    2014048319

Typeset in 10/12pt Times by SPi Global, Chennai, India.

Printed in the United States of America

10 9 8 7 6 5 4 3 2 1

1   2016

This book is Dedicated to
S G Deshmukh
For being my friend, philosopher and guide

CONTENTS

6   Two-Way Mixed Design   125

PREFACE

I got inspired to write this book when I started teaching course work to the PhD scholars. During interaction with the participants while conducting many research workshops I used to get the feedback on conducted sessions of 'research experiments'. Numerous requests for guidance came from research scholars on how to conduct research experiments that used various designs with repeated measures. Many researchers use Repeated Measures Design (RMD) in their studies due to non-availability of sufficient subjects. As a matter of fact, the curriculum of various master degree programs does not elaborate much on this technique. Hence, researchers' understanding is limited in this area. While conducting an experiment, many scholars stumble not because they have difficulty in learning a specific research design but instead they are not equipped to handle problem-solving approach in their experiments.

Although a lot of content is available on the independent measures design but very little of it deals with repeated measures design. It is the task of researchers to understand these designs and interpret their findings from such material because they do not provide in-depth explanation to the solutions. This book on RMD has been written to fill this gap. The intention of writing this book is to help the empirical researchers in any area to understand situations where such designs can be used, and provide a handy solution to analyze them with the popular IBM® SPSS® Statistics software (SPSS). Each design in the book has been discussed with solved illustrations and detailed interpretation of findings.

This book emphasizes the importance of this design in any experimental research and discusses the most widely used RMDs in empirical studies. It provides readers with basic understanding of statistical designs and facilitates them in solving and generating insights of repeated measures designs to address their research issues. The content of this book can be covered in two credit courses in the curriculum of Master's/M.Phil/PhD programs for most of the disciplines such as Psychology, Management, Social Sciences, Medicine, Physical Education, Sports Sciences, Nutrition, and so on. One of the important features of this book is that it follows a very simple approach to make the concept clear with solved illustrations.

After understanding the requirements of researchers in experimental studies it took me around five years to write this book. During the course of writing, the focus was on how to make the readers understand all the statistical designs discussed in the text. Even the most complex design like two-factor mixed MANOVA design has been discussed in a simple manner so that the researchers may be tempted to use in case of need. This book has been structured in such a manner so as to address the FAQs of the researchers as experienced by me. Each chapter has been deliberated with the finest details so as to satisfy the need of applied researchers and satisfy the theoretical statisticians as well.

The main USP of this book is that it provides comprehensive solutions along with interpretations and guidelines for using the SPSS software to the researchers. The book contains eight chapters, which are planned in increasing order of conceptual difficulty usually experienced by the researchers. The first three chapters are focused on understanding the mechanism in solving experimental designs. In chapter four to eight, various repeated measures designs have been discussed. These chapters start with the introduction of design, advantage and disadvantage of the design, application area and layout design. Finally, a solved example with the SPSS software has been illustrated in each chapter to understand the procedure and interpretation of outputs in the design.

Chapter 1 introduces fundamentals of experimental designs. It discusses various principles of design of experiments and explains all the basic statistical designs so as to build up the foundation of the readers in understanding various designs mentioned in different chapters. Various terms used in solving different designs have been discussed. A thorough discussion has been made on how to develop a good empirical study by controlling various errors in the experiment.

Chapter 2 focuses on understanding the basic mechanism in solving independent measures and repeated measures ANOVAs. Procedure involved in pair-wise comparison of means has been shown in both types of ANOVAs. An effort has been made to make the readers understand, how splitting of the total sum of squares is different in both the cases and how the repeated measures design enhances efficiency in experiments.

Chapter 3 discusses assumptions required for the repeated measures design in detail along with the methods to test them by using the SPSS software. Remedial measures in case of violating the assumptions have been deliberated in detail.

One-way repeated measures design has been discussed in Chapter 4. Different situations have been shown where this design can be used in order to help the readers to select this design for their studies. Besides this, layout of this design has been discussed in two different types of studies; firstly, in which the levels of within-subjects factor are different treatments and secondly, where they are different time durations. The steps involved in this design have been explained to provide a road map to the researchers in solving the design. Solved illustrations will help the readers to use them effectively in their studies.

Chapter 5 deals with the two-way repeated measures design. It describes how to use this design in a situation where both the factors in a factorial experiment are within-subjects. Different applications have been discussed in the chapter, and a procedure for testing various assumptions in the design has been illustrated in order to facilitate the researchers to use them in their studies. A thorough treatment has been given to analyze the main and simple effects in the design by using the SPSS software.

Chapter 6 explains the two-way mixed design in which one of the factors is a between-subject and the other is a within-subject. Detailed treatments on testing assumptions and investigating the main and simple effects have been deliberated so as to draw meaningful conclusions in the study.

Chapter 7 is devoted to the one-way repeated measures MANOVA. A detailed discussion has been made about the situations where this design can be used. There is an elaborate description as to why this design should be used in experimental studies and how to develop hypotheses based on research questions to be investigated in the studies.

Finally, in Chapter 8 we have discussed mixed design with two-way MANOVA in detail. Often researchers face difficulty in using this design; hence, an intensive work has been done to make them comfortable by creating a road map. A solved example has been discussed in detail for clarifying the procedure involved in multivariate and univariate testing. Minute details have been provided to analyze the simple effects in the univariate analysis. A thorough guideline has been provided to deal with the family-wise error rate, which inflates due to multiple univariate analysis in this design.

I hope that this book would be helpful to the researchers for their course work and research studies.

I would like to express my sincere gratitude to my friend Prof. Jagdish Prasad who has reviewed few chapters and edited the text which has enhanced the quality of the book. I am thankful to my professional colleagues namely Prof. D. P. Singh, Prof. Y. P. Gupta and Prof. V. Sekhar for helping me to edit some of the chapters in this book. I am indebted to Dr. Indu Bora for timely help in providing her expert comments in correcting some portion of the manuscript.

I would like to place on records my gratitude to Susanne Steitz-Filler, Senior Editor, John Wiley & Sons and her team for providing me all the support and encouragement in presenting this book to the audience in its present form. I am thankful to Sari Friedman for providing me all the support and timely guidance in the publication of this text. She was very cooperative and supportive in dealing all my queries during the whole process of publication. At last I would like to thank Nomita Swaminathan, Production Editor, Kiruthika Balasubramanian, Project Manager and her team in producing this book by typesetting and editing the entire text.

Thanks to all my research scholars and numerous researchers who had posed a variety of questions on research designs, which helped me in identifying the contents of this book. Above all, I want to thank my wife, Haripriya, and children Prachi and Priyam and the rest of my family, who supported and encouraged me in spite of all the time it took me away from them. It was a long and difficult journey for them.

At last I beg forgiveness from all those who have been with me over the years and whose names I have failed to mention.

For Instructors and Students Supplementary material for the book is available on a companion website, which is accessible via "www.wiley.com" at some point. I request the readers to send in their suggestions and queries to me via e-mail at vermajp@bsnl.in and I shall respond to them at the earliest.

Director, Centre for Advanced Studies   PROF. J. P. VERMA, PhD
Lakshmibai National Institute of Physical Education,
Gwalior
E-mail: vermajp@bsnl.in

ILLUSTRATION CREDITS

The IBM SPSS Statistics has been used solving various applications in different chapters of the book with the permission of the International Business Machines Corporation, © SPSS, Inc., an IBM Company. The following screen images of the software are Reprinted Courtesy of International Business Machines Corporation, © SPSS. "SPSS was acquired by IBM in October 2009."

1

FOUNDATIONS OF EXPERIMENTAL DESIGN

INTRODUCTION

Empirical research provides knowledge to the researchers through direct or indirect observations or experiences. Empirical research may either involve correlational or experimental approach. In correlational research one looks to establish relationship between two variables. In such studies a premise is made that two variables may be related in some way and then values of both the variables are obtained under different conditions to test a hypothesis if indeed there is a relationship between the two. The obtained correlation is tested for its significance. The drawback of the correlational study is that it does not establish the cause and effect relationship even if the correlation is found to be statistically significant. For instance, if the observed correlation between the caffeine intake and concentration of mind is significant and positive, it cannot be said that caffeine causes concentration. The increase in the concentration due to the increase in the caffeine intake may be due to age, motivation, gender, other lifestyle parameters.

On the other hand, experimental research provides cause and effect relationship because in such experiment a treatment is deliberately administered by a researcher on a group of individuals or objects to see its impact under a controlled environment. In other words, if changes are made in the variable A that leads to changes in variable B, one can conclude that A causes B. For example, to see the impact of exercise on muscular strength a researcher may administer different intensity of exercise to different groups of individuals to see its effect. If a particular intensity of exercise improves

Repeated Measures Design for Empirical Researchers, First Edition. J. P. Verma.
© 2016 John Wiley & Sons, Inc. Published 2016 by John Wiley & Sons, Inc.

muscular strength more than others, one may conclude that exercise intensity causes muscular strength. On the other hand, if there is no difference in the average muscular strength among different exercise groups, it may be inferred that the exercise intensity has nothing to do with muscular strength.

Authenticity in an experimental research is ensured only when an appropriate experimental design is used. Experimental design is a blueprint of the procedures which enables a researcher to test his hypothesis under a controlled environment. It describes the procedure of allocating treatments to the individuals in a sample. There are many ways in which an experimental design can be classified. One such classification is based on the method of allocating treatments to the subjects. On the basis of this criterion, experimental design can be classified into three categories; independent measures design, repeated measures design, and mixed design. In independent measures design each subject gets one and only one treatment, whereas in repeated measures design each subject is tested under all treatments. In mixed design each subject receives one and only treatment of first factor, but gets tested in all the treatments of second factor. This book specifically deals with some of the important repeated measures designs and mixed designs. To understand these designs and its applications, it is important to understand different aspects of experimental research such as principles of experimental design, types of statistical designs, terminologies used, and other considerations in planning an experimental research.

WHAT IS EXPERIMENTAL RESEARCH?

An *experimental research* is a process of studying the effect of manipulating independent variable on some dependent variable(s) observed on subjects in a controlled environment. For instance, in studying the effect of progressive relaxation on concentration, the progressive relaxation is an independent variable whereas the concentration is a dependent one. While conducting an experimental research, a researcher always tries to maintain control in an experiment so that valid conclusion can be drawn on the basis of findings. In experimental research the experimenter is allowed to manipulate independent variable to see its impact on the dependent variable. For instance, in the above example the experimenter can decide the duration or the intensity of the progressive relaxation program. Since the experimenter manipulates an independent variable to see its impact on dependent variable, cause and effect relationship can be explained on the basis of findings.

On the other hand in *observational study*, a researcher collects and analyzes data without manipulating independent variable. Here also the relationship is investigated between independent and dependent variable observed on the subjects. Since researcher is not allowed to manipulate an independent variable, causal interpretations cannot be efficiently made. If relationship is investigated between height and vertical jump performance of sprinters, the observed correlation may not be the strong evidence for causal relationship between them because the independent variable, height, has not been manipulated to see its impact on the vertical jump performance. This is because the experimenter cannot observe the control on the study. The subjects might have different weight, skill, motivation, and fitness level which do not allow interpreting the strong cause and effect relationship between height and the vertical jump performance. The observational study is also known as *correlational study* or *status study*.

Since validity of findings in an experiment depends upon the control observed during the experimentation, it is important to design the experiment in such a way so as to minimize the error involved in it. Using appropriate design in an experiment ensures proper allocation of treatments to the subjects so that experimental error is minimized. This ensures internal validity in the experiment. Design of experiment along with its principles has been discussed in detail in the following section.

DESIGN OF EXPERIMENT AND ITS PRINCIPLES

Design of experiment can be defined as a roadmap for organizing an experimental study for testing a research hypothesis in an efficient manner. Design of experiment facilitates an experimenter to observe control in an experiment, thereby reducing the experimental error and ensuring internal validity in findings. More specifically it provides a plan according to which treatments are allocated to the subjects in order to reduce experimental error. While planning a study a researcher needs to design an experiment in such a manner that the similarity is ensured among the experiential groups. The experimental error is controlled by controlling the effect of extraneous variables. To design an experiment a researcher must have the knowledge about homogeneity of experimental material or the subjects on which the experiment is required to be conducted. Besides, one should be able to identify those extraneous variables which may affect findings if not controlled. Depending upon the homogeneous conditions of subjects, an experimental design is identified. There are ways and means in testing the efficiency of design used in a research study. The efficiency of two different designs in the same experiment may be compared by using the error variance. This has been shown in Chapter 2. To have the control in an experiment and ensuring maximum accuracy in findings, Ronald A. Fisher has suggested the three basic principles of design of experiment, namely, Randomization, Replication, and Blocking.

Randomization

One of the main principles of design of experiment is randomization. *Randomization* refers to randomly allocating treatments to the subjects. Randomization ensures similarity in the experimental groups. It controls bias and extraneous variables which might affect findings of the study. Readers must note that the random selection of subjects and random allocation of treatments are two different things. Consider a study in which three different types of beverages, tea, coffee, and soft drink, are compared for their effect on reaction time. If 30 subjects are selected in the study let us see how randomization is done. Firstly, the initial sample of 30 subjects is selected randomly from the population of interest. Out of these subjects three subjects are randomly selected and the treatments are allocated randomly to them. Then another three samples are selected randomly from the remaining lot and treatments are again randomly allocated to them. In this all 30 subjects are assigned to three different treatment groups. In this study selecting 30 sample subjects randomly from the population of interest does not ensure that the treatment groups are similar, but helps the researcher to generalize findings about the population from which the sample has been drawn.

In other words, random selection of subjects ensures external validity in findings. Complete randomization is only possible if subjects are uniform. On the other hand, perfect random allocation of treatments to the subjects ensures that treatment groups are homogenous and do not contain any bias. This random allocation of treatments to the subjects ensures internal validity in findings. If external and internal validity are ensured, one can be quite sure in the above experiment that whatever the effect of a particular beverage on the subject's reaction time is observed, it is due to the beverage only and not due to any other reason. Besides this, randomization also provides the validity of F-test. Further, assumption of independence of observations in F-test also gets satisfied due to randomization.

Replication

Replication refers to repeating the treatment a number of times on different subjects. It is a fact that a treatment applied on single subject does not provide sufficient evidence of the effect of that treatment, so replication is needed. Single observation also does not provide the valid estimate of the parameters in the study so replication of treatments is essential. It is also a fact that standard error of sample mean or difference of sample means (group means) is inversely proportionate to the replication of treatments. So if number of replication increases the error variance or standard error decreases. If the treatment is effective the average effect of replication will reflect its experimental worth. If it is not few subjects in the sample who may have reacted to the treatment will be negated by the majority of subjects who were unaffected by it. If the above mentioned experiment of beverages is administered on three subjects only, and if the soft drink is found to improve the reaction time, the result may not be acceptable until unless this result is observed on most of the subjects in the sample. Thus, replication reduces variability in experimental results and provides confidence to the researcher in drawing conclusion about the effect of treatment on dependent variable.

Blocking

Blocking is a technique of reducing experimental error by including an extraneous variable in the experiment. Blocking refers to dividing heterogeneous experimental units into homogenous blocks so that the units in the blocks are homogeneous. In other words whole experimental material is divided into homogeneous strata. Blocks are made if experimenter has some knowledge about the experimental material or subjects prior to conducting an experiment, through pilot study or uniformity trial or some prior studies. Blocking technique is used if an experimenter knows that the variability exists among the subjects. Generally, size of the block should be equal to the number of treatments. After dividing the experimental units into blocks, the treatments are randomly allocated in each block. Blocking enhances precision in the study by reducing the experimental error. In the above example of beverages if gender is considered as blocking variable, the blocks of male and female may be made in which treatments may be allocated randomly.

STATISTICAL DESIGNS

It is important for the researchers to know about various kinds of statistical designs so as to choose the appropriate one for their study to obtain reliable findings. Selection of design depends upon many parameters such as number of factors to be investigated, variability of the experimental units, and degrees of precision required. In empirical research, studies can be classified in two categories; single factor studies and multi-factor studies. In single factor study the effect of only one factor on some dependent variable is investigated, whereas, in multifactor studies effect of two or more independent factors on some dependent variable is investigated. Multifactor studies are also known as factorial experiment. Factorial experiment may have two or more independent factors, each having two or more levels. All these studies can be conducted by using any of the three basic statistical designs namely; Completely Randomized Design, Randomized Block Design and Latin Square Design. Choice of using these designs depends upon the knowledge about variability of the experimental materials or subjects. We shall discuss these designs briefly later in this chapter.

In the example of the beverages discussed above the effect of only one factor is investigated, hence the design used for analysis would be one factor design. Here the treatment factor has three levels and therefore three treatment conditions are to be compared for their effectiveness. But if along with the beverage, the effect of duration is also required to be investigated, the experiment is said to be the two factor(or multifactor) study. Similarly, more than two factors can also be simultaneously investigated and the design in such situation would be known as multifactor design. If the effect of more than two factors is investigated simultaneously, the analysis becomes very complex and therefore researchers usually investigate the effect of either one or two factors only. Whatever design a researcher uses, it is analyzed by using the group of techniques known as analysis of variance (ANOVA). In any experimental design the three basic principles of randomization, replication and blocking are used. Selection of the design of experiment depends upon the fact as to how we wish to carry out these principles. All types of designs discussed above will be explained with the help of examples in the following sections.

Completely Randomized Design

Completely randomized design is the simplest design used by a researcher to test the effectiveness of one factor (with two or more levels) on some dependent variable. This design is used when the entire experimental units (or subjects) are homogeneous in all respect and the experimenter has full control on the experiment. Under such strict control whatever the effect of the independent variable is observed on the dependent variable, can be totally attributed due to the independent variable only. This kind of experiment is also known as laboratory experiment. In completely randomized design

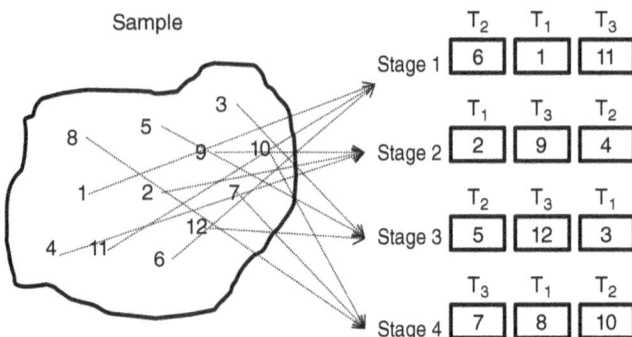

Figure 1.1   Layout of the completely randomized design

sample size in different treatment groups may differ. In this design treatments are randomly allocated to the subjects.

If the effect of three treatments, T_1, T_2, and T_3, are to be compared for their effectiveness on some dependent variable by administering them on a randomly drawn sample of size n, let us see how the treatments are allocated to the subjects in a completely randomized design. In the first stage, three subjects are randomly selected from the sample and all the three treatments are randomly allocated to these subjects. Then again three subjects are randomly selected from the remaining subjects in stage 2 and the treatments are again randomly allocated to these three samples. This process will go on till all the sample units are allocated in one or the other treatment groups. The readers must note that in this design each sample unit will get one and only one treatment. The layout of this design in which allocation of three treatments on 12 sample units has been done is shown in Figure 1.1. Randomization of treatments to the sample units ensures internal validity in the study.

Consider an experiment in which the effect of music is to be seen on concentration. The researcher may have three different types of music; classical, instrumental, and orchestra. This experiment can be organized by using completely randomized design in which the treatments can be randomly allocated to the samples as discussed above.

Randomized Block Design

Randomized block design (RBD) is used in a situation where the experimental units (or subjects) are not homogeneous. The reason of non-homogeneity of experimental units is known to the experimenter before applying the RBD. This reason may also be called classification or categorical variable not the treatment. Levels of these variables are also known to the experimenter. In using this design the entire sample is divided into blocks of homogeneous subjects and then the treatments are randomly allocated to these subjects within each block. Here the block is known as classificatory variable. The experimenter divides the sample into as many blocks as the levels of categorical variable. In a randomized block design, if m treatments are to be replicated r times then r blocks need to be created and $m \times r$ subjects need to be selected in the

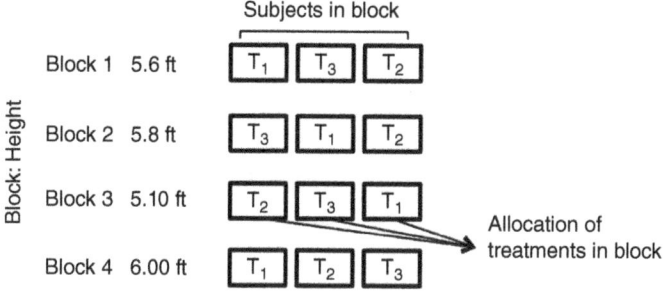

Figure 1.2   Layout of the randomized block design

study. Each of the *m* treatments is assigned randomly to only one subject in each block. The random allocation of treatment in each block is done independent of other blocks. The blocks are made on that variable which is supposed to affect dependent variable. Consider a study in which the effect of three different types of teaching methodology is compared to see their effect on the improvement of the subject's learning efficiency. It is a known fact that learning efficiency depends upon IQ level of the subjects. Thus, if the subjects vary in their intelligence level, the blocks may be made on IQ. For instance if this experiment is conducted on 12 subjects to compare the effectiveness of three teaching methodology (T_1, T_2, and T_3), the subjects may be divided into four blocks of different IQ levels, having three subjects in each. All the three treatments will then be randomly allocated in each block. The layout for 12 subjects, grouped in 4 blocks, can be shown by the Figure 1.2. Here the blocks have been made on the basis of the IQ. The reader should note that the blocks must be homogeneous within itself and heterogeneous among themselves in relation to IQ.

Another variant of the randomized block design is the one in which treatments are replicated many times within each block. For instance, if the above experiment is conducted by making the block on gender instead of IQ, all the three treatments T_1, T_2, and T_3 would be replicated many times in each gender. For instance, if a study is conducted to compare the effect of these teaching methodologies on 120 subjects, 60 male and 60 female would be taken as the subjects in the study. The treatments T_1, T_2, and T_3 will then be randomly replicated 20 times in male as well as in female group. The reader should note that in each block the number of subjects should be in the multiple of treatments. The randomized block design is generally used in a situation where the performance of the subjects is known to vary with the variation in certain variable. For instance, performance may vary with age, height, gender, and so on. Hence, blocking can be made on such variables. This design is mostly used by the researcher because of its flexibility and robustness. However, it becomes less efficient as the number of treatments increases because in that situation the block size also increases. Hence, it is difficult to maintain the homogeneity in the blocks. This design is more efficient than the completely randomized design because experimental error is reduced due to blocking. A detailed comparison of these two designs has been shown in Chapter 2.

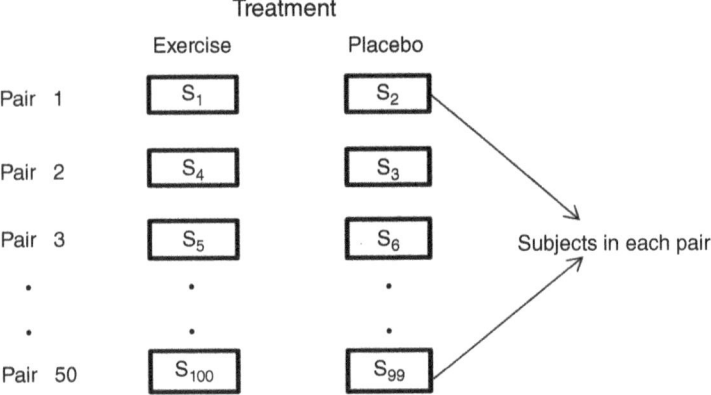

Figure 1.3   Layout of the matched pairs design

Matched Pairs Design   Matched pairs design is a special case of the randomized block design. In this design the subjects are matched on some characteristics which are supposed to affect the experiment. For instance, if the effectiveness of exercise is to be investigated, matching of subjects should be done on the basis of gender and age because these variables affect the experiment, but if the matching is done on the basis of IQ no benefits would be derived from this design. This design can be used to compare only two treatments. The advantage of this design is that it can explicitly control more than one extraneous variable, for instance the age and gender in the above mentioned example. Here each matched pair is like a block. Consider an experiment in which effect of exercise on strength is to be studied on 100 students. In using this design we shall divide these 100 subjects into 50 pairs of subjects on the basis of gender and age. For instance, pair P1 may have both the subjects as female with 19 years of age and the pair P2 may have both the subjects as male with age 21 years. In each pair treatments are randomized. If one of the treatments is a low intensity exercise developed by the researcher and the other is a placebo, either of the two subjects may receive low intensity exercise and the other may receive placebo. Similarly treatments are randomized in each pair separately. The layout of this design can be understood by the Figure 1.3.

The matched pairs design is superior to the completely randomized design as well as randomized block design because here group becomes more homogeneous due to matching but the drawback is that it can compare only two treatments.

Latin Square designs

Latin square design is useful in a situation where the variability in the experimental material exist due to two extraneous variables in the experiment. The knowledge about the variability must be known to the researcher in advance for using this design. Since in RBD, only one extraneous variable affects the dependent variable hence we included one blocking variable in the experiment but in this design since effect of two extraneous variables needs to be controlled hence the experimental material is

Block: IQ	Block: Age — Teens	Mid age	Old age
Low	T_1	T_2	T_3
Average	T_3	T_1	T_2
High	T_2	T_3	T_1

Figure 1.4 Layout of Latin square design

divided in two different blocks (rows and columns). Thus in this design two blocking variables are taken. Consider a study in which the effect of three different types of teaching methodology T1, T2 and T3 on learning efficiency needs to be investigated. If it is known to the researcher that the learning efficiency in this experiment is known to be affected by the IQ and age of the subjects then the subjects would be divided in row(IQ) and column(Age) blocks. One of the restrictions in this design is that one needs to have equal number of blocks on both the extraneous variables. Number of treatments should also be equal to the number of blocks of each variable. Let us take three IQ and three Age groups in this study. Thus, three treatments can be taken in this study which can be replicated on three subjects. While allocating treatments to the subjects, each treatment should occur exactly once in each row and exactly once in each column. If T1, T2 and T3 represent traditional teaching, audio-visual teaching and flexible teaching methods respectively then one of the layout in LSD can be as shown in Figure 1.4. The main advantage of this design is that it requires less number of subjects to investigate different treatments. For instance, in this experiment 27(3×3×3) treatments can be compared by using only nine subjects.

FACTORIAL EXPERIMENT

If we consider more than one manipulated variables at different levels and their interaction effects are also to be judged then the factorial experiment is considered. Factorial experiment is used to investigate the effect of more than one factor at two or more levels on the dependent variable simultaneously. The factorial experiment is represented by m^p, where "p" represents the number of factors and m indicates the number of levels of each factor has. Thus, 2^2 factorial experiment represents that there are two factors each having two levels, and in all, have four treatment combinations. If a factorial experiment has $p+r$ independent factors so that each of the p factors has m levels and each of the r factors has n levels then the experiment is represented as $m^p \times n^r$ factorial experiment. For instance, if the factorial experiment has two factors A and B having levels 2 and 3, respectively, then it is called 2×3 factorial experiment.

Besides investigating the effect of each factor on the dependent variable, one can simultaneously investigate the interaction effect (joint effect) of the two factors on the dependent variable also. Thus, in factorial experiment all the levels of one factor may be compared in each level of the other factor. The simplest factorial experiment is 2×2, in which each factor has two levels. If the factor A has two levels (A_1, A_2) and the factor B has three levels (B_1, B_2, and B_3), this experiment is known as 2×3 factorial experiment. Thus, in this experiment six treatment combinations (A_1B_1, A_1B_2, A_1B_3, A_2B_1, A_2B_2, A_2B_3) need to be compared and therefore six groups of random samples need to be selected. If experimenter decides to replicate each treatment condition on five subjects, $30 (= 5 \times 2 \times 3)$ subjects need to be randomly selected from the population of interest. This experiment may be conducted in CRD, RBD or LSD designs as per the requirement of experiment. If the factorial experiment is conducted in CRD then all six treatment conditions shall be randomly allocated to these 30 subjects in sample so that each treatment is received by five subjects thus making five experimental groups. On the other hand due to heterogeneity considerations, if the factorial experiment is using RBD then these six treatment conditions shall be randomly allocated in each block. In this experiment the main effect of both the factors A and B can be investigated along with the simple effect of each factor simultaneously. The main effect can be defined as the effect of first independent variable (Factor A) on the dependent variable across all the levels of the second independent variable (Factor B). The interaction is ignored for this part. Just the rows or just the columns are used, not mixed. This is the part which is similar to one-way analysis of variance. Each of the variances calculated to analyze the main effects (Rows and Columns) is like between variances.

The advantage of factorial experiment is that one can test the significance of interaction between the two or more factors. The interaction can be defined as the joint effect of two factors or more on the dependent variable. It can also be defined as the effect that one factor has on the other factor.

In factorial experiment if the interaction effect is not significant, the main effect becomes meaningful. On the other hand, if the interaction effect is significant, the simple effects need to be investigated. The simple effect is the effect of one independent variable on the dependent variable in each level of the other factor. It is investigated by comparing the effect of all the levels of one factor in each level of the other factor.

Consider an experiment in which the effect of mental exercise (A) with three different intensities (low, medium, and high) and environment(B) with three different climatic conditions (hot, humid, and cold) are to be investigated to see their impact on the task efficiency of the subjects. This study can be conducted as a 3×3 factorial experiment organized using any of the three designs namely CRD, RBD or LSD depending upon whether the subjects in the sample have no variability, one dimensional variability or two dimensional variability respectively. Thus, in this design nine treatment combinations need to be compared. If an experimenter decides to replicate each treatment on five subjects using CRD then this experiment is called as 3^2 factorial experiment conducted in CRD with five replications. The layout design of this experiment can be shown by the Figure 1.5.

TERMINOLOGIES IN DESIGN OF EXPERIMENT

Understanding the following terminologies shall facilitate the readers to learn different designs discussed in this book in a better way.

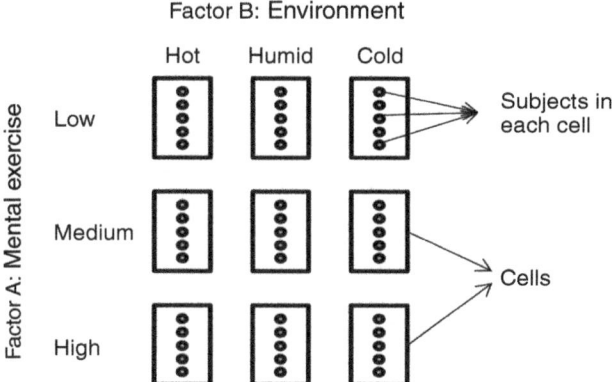

Figure 1.5 Layout of the 3×3 factorial design

Subject

Subject is usually used for an individual on whom an experiment is conducted. In investigating the effect of incentives on buying behavior, the individuals to which the incentives are offered are known as subjects.

Experimental Unit

Experimental unit can be defined as the subject/object on which the treatment is to be administered in an experiment. If the effect of beverages is investigated on concentration, the subjects consuming beverages are known as experimental units. Similarly if an impact of government policy is to be studied on the sports facilities in schools, the school would be an experimental unit.

Factor and Treatment

Factor can be defined as the independent variable whose effect is to be seen on the dependent variable whereas, different levels of factors are known as treatments. For instance, in beverages experiment, beverage is known as factor and different types of beverages are referred as treatments. In experimental research a researcher is allowed to manipulate the independent variable and therefore if one decides to compare the effect of three different types of beverages, say beer, coffee, and tea, on concentration, the beverages is said to have three levels or three *treatments*. The beverages will be known as *Factor* or *independent variable*. In other words, the *level* of a factor is the number of variation in the independent variable which an experimenter wishes to compare in relation to the response on the dependent variable.

In investigating the effect of independent variable on some dependent variable where the independent variable has not been manipulated by the experimenter is known as *classificatory variable*. For instance, to see the effect of gender on IQ among the group of students, the gender is a classificatory variable because gender has not been manipulated by the experimenter and the subjects have been classified on the basis of their possessed characteristics.

Criterion Variable Criterion variable can be defined as a variable in which the investigator is interested to see as to how it behaves due to manipulation in the independent variable. It is usually a dependent variable in the study. The criterion variable is also known as response variable. There can be many criterion variables in the study. Consider an experiment in which an investigator wishes to investigate the effect of chocolate type on its taste. Here taste is a criterion variable which is of prime importance to the researcher. The investigator manipulates chocolate by having its different variants say white, milk, and dark whose effects may be compared on taste. Thus, in this case taste depends upon different types of chocolate. Further, in this experiment if the investigator is also interested to see the effect of chocolate type on crunchiness along with the taste, there would be two criterion variables, that is, taste and crunchiness. Similarly depending upon the objectives of the study the researcher may have several criterion variables in the study. Such study is a case of multivariate analysis of variance.

Variation and Variance

Variation refers to the spread of scores around mean values, whereas the variance can be defined as a measure of variation. Variance is simply the square of standard deviation and is given by the following formula:

$$\sigma^2 = \frac{1}{n} \sum (x - \mu)^2 \tag{1.1}$$

This σ^2 is the population variance and can be estimated by the sample mean square, S^2, as given by the formula (1.2).

$$S^2 = \frac{1}{n-1} \sum (x - \bar{x})^2 \tag{1.2}$$

The estimate of variance due to independent variable(s) and variance due to error are estimated to solve different designs. Detail procedure has been shown in Chapter 2.

Experimental Error

Experimental error can be defined as an error which cannot be controlled in an experiment. Mostly this is the result of individual variation during an experiment. One cannot attribute the reason for such variation. By having control on the experiment and using proper experimental design, the researcher tries to reduce it. For instance, in comparing the effect of three different chocolates on taste, the observed effect on the subject's taste cannot be attributed to the chocolate type alone but it is partially due to the subject's characteristics such as age, gender and socioeconomic status as well. This variation which cannot be accounted for due to the chocolate is known as experimental error.

External Validity

External validity refers to the extent of generalizibility of research findings to the population from which the sample is derived. To ensure the external validity, it is important that the sample is randomly drawn from the population of interest. Thus, in a study if the sample is randomly drawn by using an appropriate probability sampling technique, external validity can be ensured. On the other hand, if the sample is drawn by using any of the nonprobability sampling method, the study lacks external validity.

Internal Validity

In an experimental research if the effect on a criterion variable can be attributed due to manipulation in an independent variable, the research study is said to have internal validity. Internal validity refers to the extent of which one can say that the variation observed in the dependent variable is due to the variation in the independent variable. To ensure the internal validity, the external variance should be controlled in the experiment. The randomization of experimental units to the subjects is one of the best ways to ensure the internal validity in the study.

CONSIDERATIONS IN DESIGNING AN EXPERIMENT

We have seen that in experimental research a researcher is interested to see the impact of some independent variable or factor on the criterion variable. A *factor* which can be manipulated by the researcher to see its impact on the criterion variable is known as treatment variable. Examples of treatment variables are exercise program, nutrition, pranayama, training intensity, incentives, and so on. In all such cases an experimenter can manipulate them. A treatment variable is basically an independent variable. If the independent variable cannot be manipulated by the researcher, it is known as classificatory variable. The *classificatory variable* can be defined as some preexisting characteristics of the subjects on the basis of which they can be classified. The variables like gender, socio economic status, sports category, geographical region, and weather conditions are all classificatory variables because these variables cannot be manipulated by the researcher. For instance, in seeing the impact of three different types of nutritional supplements on the muscular strength, the nutritional supplement is a treatment variable because it can be manipulated by the experimenter by choosing its different types. However, in investigating the effect of gender on the IQ, gender is a classificatory variable because the gender cannot be manipulated by the experimenter.

In two-factor design, in which the experimenter is interested to see the impact of two factors simultaneously on some criterion variable, one of the factors can be treatment variable whereas the other can be a classificatory variable. Consider an experiment in which the effect of sleep deprivation (with three different durations) is to be seen on the task performance in male and female subjects. In this case the two independent variables are sleep deprivation and gender, whereas the criterion variable is the task performance. Out of these two independent variables the sleep deprivation is a treatment variable because the experimenter can manipulate its duration, whereas the gender is a classificatory variable because it cannot be manipulated as the subjects are classified on the basis of their gender, a preexisting criterion among the subjects prior to conducting the experiment.

Most of the time a researcher manipulates an independent variable in his experiment because by doing this more control can be observed in the experiment resulting more accuracy in the findings. Manipulating independent variable depends upon the nature of the study. Sometime the objective of the study is such that the independent variable cannot be manipulated by the researcher, in that case the independent variable is taken as a classificatory variable. For instance, if the effect of consuming tobacco on the athlete's performance is to be investigated, the subjects can be classified into nontobacco, occasional user, and regular user categories. In this case independent variable cannot be manipulated by the researcher as the subjects cannot be forced to consume tobacco to see its impact on their athletic performance. In fact it is unethical if somebody does that.

Choosing statistical design depends upon the objective of the study and the knowledge about the heterogeneity of experimental material (subjects). An experimenter is required to manage different types of variances in such a way in the experiment so as to get the reliable findings. The purpose of each and every design is to reduce the error in the experiment for enhancing the reliability of findings. This is done by maximizing the systematic variance, controlling the extraneous variance, and minimizing the error variance in the study (Verma, 2014). By following these guidelines reliable findings in the experimental study can be achieved. Detailed discussions about these aspects have been made in the following sections.

Systematic Variance

Systematic variance refers to the measure of variation resulting from manipulating the independent variable by the researcher. The researcher must choose different levels of the independent variable in such a manner so as to have the maximum variability in the criterion variable. For instance, in the above mentioned sleep deprivation experiment the researcher must choose the duration in such a way so as to achieve the maximum variance in the task performance of the subjects in different treatment groups. Since the researcher has a freedom to manipulate this independent variable, he might choose to have three durations of sleep deprivation as 24 hours, 30 hours, and 36 hours. Of course choosing these durations must be based on some scientific theory or literature. However, if an experimenter decides to have these three durations as 22 hours, 23 hours, and 24 hours, this may not produce maximum variance in the criterion variable during experiment. Thus, a researcher must use this opportunity to maximize the systematic variance by scientifically choosing the levels of the treatment variable(s).

Extraneous Variance

While investigating the effect of independent variable on some criterion variable the researcher must ensure that whatever the variation is observed in the criterion variable, most of it can be explained due to the manipulation of the independent variable. This can be achieved by controlling the effect of extraneous variables and making the experimental groups as similar as possible. This can be done by using any one or more methods discussed below.

Randomization Method   Random allocation of treatments to the subjects is the most powerful method of controlling the external variance in the experiment. It ensures the effect of all external variables to be equally distributed in the treatment groups. Randomization ensures similarity of all the treatment groups. This ensures internal validity by removing the confounding effect of the extraneous variables in the findings of a research experiment. In fact randomization is one of the main principles of the design of experiment. Randomization also controls bias in allocating treatments to the subjects. Another advantage of randomization is that it ensures normality of data in different groups which is one of the main assumptions in solving the design by using the analysis of variance. Thus, a researcher should not only draw the random sample from the population of interest, but should also allocate the treatments randomly to the subjects for controlling the external variance in the experiment.

Elimination Method   Elimination method is another way of controlling the external variance in the experiment. In this method the extraneous variable which affects the criterion variable during experiment is stabilized in the study. Let us consider an experiment in which the effect of three different exercise programs is to be compared on weight reduction. Let us further assume that these programs are administered on three different groups of subjects which include male and female both. In this case weight reduction cannot be solely due to the variation in the exercise programs, but partly affected by the gender as well. In order to done away the effect of gender, the whole experiment can be organized either on male or female. In other words, effect of gender can be controlled by eliminating either of the two genders from the study. The problem with such study is that the findings cannot be generalized for the entire population. In this case if the female subjects are eliminated, the findings can only be applicable to the male subjects and not for the females; hence, elimination method lacks in external validity. Thus, if a researcher has an idea about the extraneous variable which affects criterion variable in an experiment, the elimination method may control the external variance provided objective of the study demands for it.

Matching Group Method   Another method of controlling external variance is by means of matching the group on some extraneous variable. If the effect of three training programs (T_1, T_2, and T_3) is to be seen on the shooting accuracy in basket ball, the subjects may be matched in the groups on the basis of their height because height is supposed to affect the shooting accuracy besides training programs. Using this method the height of all the subjects in the sample is measured first and scores so obtained are arranged either in ascending or descending order. After arranging the subjects according to their height, the first three subjects may be selected and the three treatments (in this case) can be randomly allocated to them. Again the next three subjects are selected and all the three treatments are randomly allocated to them. This way all the subjects are allocated to some or other treatment group. The readers must note that each subject will be classified into one of the three treatment groups. If the random sample consists of 12 subjects, matching of these samples on the basis of their height can be shown by the Figure 1.6.

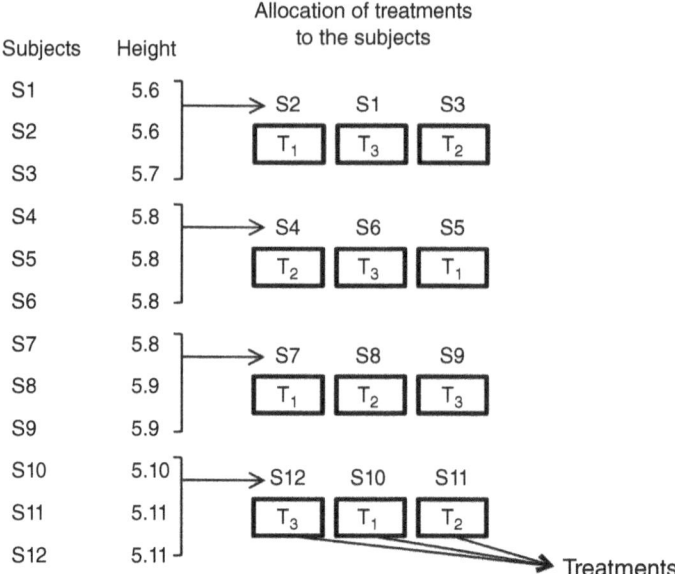

Figure 1.6 Allocation of treatments by matching the subjects

Adding Additional Independent Variable The above discussed three methods of controlling extraneous variance, that is, randomization, elimination, and matching, are known as nonstatistical methods. If the variability of an extraneous variable cannot be controlled by any of the nonstatistical methods, one may include it in the design. For instance, if an experimenter is interested to compare the effect of three different conditioning programs (T_1, T_2, and T_3) on the general fitness among the college students, in that case if the experimental groups consist of male and female both, gender can be included as an additional independent variable in the design. The gender is included in the design as an additional independent variable because it is considered to affect the criterion variable, that is, general fitness during experimentation. This variable is known as blocking variable. The purpose of adding gender as an independent variable is only to reduce the error variance in the design, as some part of the error variance can be explained by the variable gender. Besides reducing the error variance, this independent variable does not serve any other purpose in the experiment. The reader must note that by the inclusion of extraneous variable in the design the error variance is reduced but at the same time degrees of freedom of error variance also gets reduced. Thus, the design will become more efficient only when the extraneous variable is known to affect the criterion variable significantly. If the three treatments in the above mentioned experiment are different intensities of the conditioning program, that is, low, medium, and high, the layout of the design can be shown by the Figure 1.7.

Statistical Control We have seen that if nonstatistical procedures are not sufficient to control the extraneous variance, the statistical procedure can be adopted by including the extraneous variable in the design. Other statistical procedures through

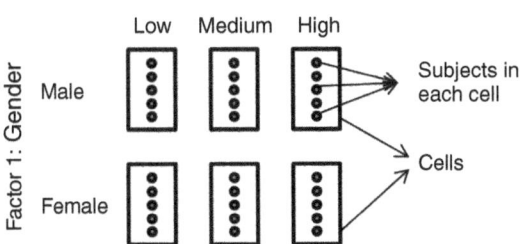

Figure 1.7 Layout of the design after including extraneous variable in the design

which control can be observed in the experiment are repeated measure designs and analysis of covariance (ANCOVA) design. In using the repeated measures design subjects serve their own control; hence, error variance is reduced to a great extent. On the other hand, in ANCOVA design the extraneous variance is controlled by adjusting the effect on the criterion variable due to treatments in relation to the extraneous variable. In ANCOVA design the extraneous variable is also known as covariate. Since this book is specifically meant for the repeated measures designs, various repeated measures designs have been discussed in different chapters along with their solutions by using the SPSS software. The ANCOVA design is out of scope of this book. The readers are advised to refer to the book Verma (2011) for this topic.

Error Variance

All nonstatistical and statistical methods discussed above to control the extraneous variance help in reducing the error variance also. Besides this by following some more guidelines the error variance can further be minimized in the design. For instance, the researcher must use the standard equipments so as to ensure the instrument reliability in data collection. Similarly the researcher must follow the proper instructions applicable in data collection of different parameters and if the study is of survey type, a clear-cut instruction must be given to the field investigators for reliable data collection.

Maximizing systematic variance, controlling extraneous variance, and minimizing error variance facilitates an investigator to compare systematic variance against error variance. Minimizing error variance gives the systematic variance a chance to show its significance. Managing all types of variances ensures that whatever the effect observed in the criterion variable is genuinely due to the treatment variable.

EXERCISE

1.1. What happens if the principles of design of experiments are not followed in conducting an experimental study?

1.2. Explain the concept of external and internal validity in an experimental study by means of examples.

1.3. What are the benefits of randomization in a research design? Explain the difference between random selection of sample and random allocation of treatments to the subjects.

1.4. What is factorial experiment? Explain its advantage in comparison to that of single factor study.

1.5. Why randomized block design is considered to be the superior design in comparison to that of completely randomized design. Discuss the layout of these two designs.

1.6. Explain the matched pairs design and discuss its advantage over completely randomized design and randomized block design.

1.7. What are the different ways and means in controlling extraneous variance in an experimental study?

1.8. In an experimental study, how different variances should be controlled? What happens if an experimenter optimally maximizes systematic variance, controls extraneous variance, and minimizes error variance?

ASSIGNMENT

1.1. A researcher decides to organize a completely randomized design to investigate the effect of different types of beverages on the concentration. He has taken three different types of beverages and a sample of 15 subjects. Describe the layout of the design in the study.

1.2. An investigator wishes to compare the effect of three different types of exercises on flexibility. He has selected nine male and nine female subjects in the study. Theory suggests that the experimental results would be affected by the gender difference. Which design you would prefer. Show the layout design in the study.

1.3. By organizing a matched pairs design an investigator wishes to see the impact of garlic on the cholesterol level of the subjects. He feels that the age and activity levels of the subjects will affect the findings in the study; hence, took these two criteria for matching the group. Suggest a plan of study and show the layout design.

BIBLIOGRAPHY

Addelman S. The generalized randomized block design. Am Stat 1969;23(4):35–36. DOI: 10.2307/2681737.JSTOR 2681737.

Addelman S. Variability of treatments and experimental units in the design and analysis of experiments. J Am Stat Assoc 1970;65(331):1095–1108. DOI: 10.2307/2284277. JSTOR 2284277.

Ader HJ, Hand DJ. *Advising on Research Methods: A Consultant's Companion*. Johannes van Kessel Publishing. 2008.

Anderson DW, Kish L, Cornell RG. On stratification, grouping and matching. Scand J Stat 1980;7(2):61–66. JSTOR 4615774.

Arceneaux K, Gerber AS, Green DP. A cautionary note on the use of matching to estimate causal effects: an empirical example comparing matching estimates to an experimental benchmark. Sociol Method Res 2010;39(2):256–282. DOI: 10.1177/0049124110378098.

Bisgaard S. Must a process be in statistical control before conducting designed experiments? Qual EngASQ 2008;20(2):143–176.

Box F. (1978) R. A. Fisher: The Life of a Scientist, New York: Wiley, http://en.wikipedia. org/wiki/Special:BookSources/0471093009 ISBN 0-471-09300-9.

Dehejia RH, Wahba S. Causal effects in nonexperimental studies: reevaluating the evaluation of training programs. J Am Stat Assoc 1999;94(448):1053–1062. DOI: 10.1080/01621459.1999.10473858.

Dehue T. Deception, efficiency, and random groups: psychology and the gradual origination of the random group design. Isis 1997;88(4):653–673. DOI: 10.1086/383850PMID 9519574.

DiNardo J. Natural experiments and quasi-natural experiments. In: Durlauf SN, Blume LE, editors. *The New Palgrave Dictionary of Economics*. 2nd ed. Palgrave Macmillan; 2008. DOI: 10.1057/9780230226203.1162.

Dunning T. *Natural Experiments in the Social Sciences: A Design-Based Approach*. Cambridge University Press; 2012.

Gates CE. What really is experimental error in block designs? Am Stat 1995;49(4):362–363. DOI: 10.2307/2684574JSTOR 2684574.

Goodwin CJ. *Research in Psychology: Methods and Design*. USA: John Wiley & Sons, Inc.; 2005.

Hacking I. Telepathy: origins of randomization in experimental design. Isis 1988; 79(3):427–451. DOI: 10.1086/354775JSTOR 234674.MR 1013489.

Heitink G. *Practical Theology: History, Theory, Action Domains: Manual for Practical Theology*. Grand Rapids, MI: Wm. B. Eerdmans Publishing; 1999. p 233. ISBN: 9780802842947.

Kempthorne O. *The Design and Analysis of Experiments (Corrected Reprint of (1952) Wiley ed.)*. Robert E. Krieger; 1979. ISBN: 0-88275-105-0.

McLeod, S. A. (2007). Experimental Design. Retrieved from http://www.simply psychology.org/experimental-designs.html.

Montgomery D. *Design and Analysis of Experiments*. 8th ed. Hoboken, NJ: John Wiley & Sons, Inc; 2013. ISBN: 9781118146927.

Moore DS, Notz WI. *Statistics: Concepts and Controversies*. 6th ed. New York: W.H. Freemanpp. Chapter 7: Data ethics; 2006. ISBN: 9780716786368.

Moore DS, McCabe GP, Craig BA. *Introduction to the Practice of Statistics*. 6th ed. New York: W. H. Freeman and Company; 2009.

Rubin DB. Matching to remove bias in observational studies. Biometrics 1973;29(1):159–183. DOI: 10.2307/2529684. JSTOR 2529684.

Verma, J. P. (2011). Statistical methods for sports and physical education. Tata McGraw Hill Education Private Limited.

Verma, J. P. (2014). Statistics for Exercise Science and Health with Microsoft Office Excel. John Wiley & Sons.

Wilk MB. The randomization analysis of a generalized randomized block design. Biometrika 1955;42(1–2):70–79. JSTOR 2333423.

2

ANALYSIS OF VARIANCE AND REPEATED MEASURES DESIGN

INTRODUCTION

One of the main advantages of experimental research is that it ensures internal validity in findings. In other words, an investigator can be sure enough in concluding that the effect observed in a criterion variable is due to the change in the treatment variable. It is because experimenter can manipulate a treatment variable in experimental studies to see its impact on a criterion variable under a controlled environment. For instance, to see the impact of caffeine on memory retention, a researcher can manipulate the level of caffeine as low, medium, and high. By allocating these three treatments randomly to the subjects ensures more internal validity of the findings as the experimenter is in full control of the experiment. On the other hand, status studies, where the independent variable is a classificatory variable which is not manipulated by the experimenter, lacks in internal validity because the subjects are classified on the basis of their preexisting characteristics. Since in such studies an experimenter cannot manipulate the independent variable, one has less control in the study. For instance, in investigating the effect of locality on the frustration of subjects the independent variable is a classificatory variable where the subjects can be randomly drawn from the rural and urban categories. Here independent variable is not being manipulated by the experimenter; hence, subjects cannot be controlled in these two groups, resulting in reduced internal validity in the study. To enhance internal validity a large number of subjects may be required so that the subjects in each category may be more representative. Thus, experimental study is the best approach to test a hypothesis. In

Repeated Measures Design for Empirical Researchers, First Edition. J. P. Verma.
© 2016 John Wiley & Sons, Inc. Published 2016 by John Wiley & Sons, Inc.

conducting an experimental study the researcher needs to design an experiment. The main issue in designing an experiment is the procedure of allocating treatments to the subjects. Based on this criterion all designs can be classified into one of the three categories; independent measures designs, repeated measures designs, and mixed design. We will now discuss these designs in short.

In independent measures design each subject receives one and only one treatment. In other words, different subjects are used in each treatment group. The allocation of subjects in different treatment groups has been thoroughly discussed in Chapter 1. The independent measures design is also known as *between-subjects design*. On the other hand, in repeated measures design same subjects are tested in each treatment condition. This means that the same set of subjects receives all the treatments. Since the same subjects are tested repeatedly, these designs are known as repeated measures design. The repeated measures design is also known as *within-subjects design.*

Repeated measures design is more convenient to apply in comparison to that of independent measures design because it requires less number of subjects and less time to conduct an experiment. The best part of the repeated measures design is that the subjects serve their own control; hence, the reduction of error variance is more in comparison to that of independent measures design. The repeated measures designs do have disadvantages as well. In such design there may be an order effect. Order effect refers to the improvement or decline in performance during testing which may be due to learning effect, or fatigue by undergoing treatments in a specific order. The performance of subjects in the second treatment condition may be better or worse in comparison to the first condition depending upon whether there exists learning effect or fatigue in the first treatment condition. Besides order effect, subjects may get bored as they undergo all the treatment conditions due to which their optimal performance may not be observed.

On the other hand, in independent measures design there is no order effect because subjects are different in different treatment conditions. The disadvantage of the independent measures design is that it requires more number of subjects; hence, their varied background and individual variations in their performance may add to error variance.

Since repeated measures design is solved by using the similar concept of analysis of variance (ANOVA) used in case of independent measures, the computation of one-way ANOVA has been shown first and then detailed computation involved in repeated measures design shall be discussed later in this chapter.

UNDERSTANDING VARIANCE AND SUM OF SQUARES

All kinds of designs in independent measures as well as repeated measures are solved by using the concept of ANOVA. It is therefore important to know the terms used in the computation of ANOVA. In solving any experimental design we deal with different kinds of sum of squares and variances, hence let us understand these terms first. The variation is the spread of scores and the variance is an index to measure the variation. In analyzing any design different types of variances are estimated by computing various sum of squares. The total variance in the population is estimated by computing the mean square variance S^2 using sample data.

$$S^2 = \frac{1}{n-1} \sum (x - \bar{x})^2 \tag{2.1}$$

where n is the sample size

$$\Rightarrow \qquad S^2 = \frac{SS}{df}$$

$$= \frac{\text{Variation}}{df}$$

$$= \text{MSS} \tag{2.2}$$

where SS represent the sum of squares and is equal to $\sum (x - \bar{x})^2$, and df denotes degrees of freedom for computing mean square variance S^2. In fact the sum of squares indicates the variation of scores around mean value, whereas mean sum of squares is an index to measure such variation. In ANOVA the term MSS is used to represent the estimated variance.

Degrees of freedom can be defined as the number of independent pieces of information that is used in the final calculation of a statistic. In ANOVA mean sum of squares are calculated by using degrees of freedom. Larger the degrees of freedom better is the estimate of variance.

Let us see how to compute sum of squares for the data 5, 6, 4, 8, 2 (Table 2.1). It can be computed by solving the expression $\sum (x - \bar{x})^2$ as shown below:

$$\begin{aligned}
SS = \sum (x - \bar{x})^2 &= \sum (x^2 + \bar{x}^2 - 2.x.\bar{x}) \\
&= \sum x^2 + \sum \bar{x}^2 - 2 \sum x.\bar{x} \\
&= \sum x^2 + N\bar{x}^2 - 2N.\bar{x}^2 \qquad \left(\sum x = N\bar{X} \right) \\
&= \sum x^2 - N.\bar{x}^2 \tag{2.3}
\end{aligned}$$

Thus,

$$SS = \sum x^2 - \frac{G^2}{N} \tag{2.4}$$

Table 2.1   Computation of Sum of Squares and Mean Sum of Squares

X	X^2	$(X - \bar{X})$	$(X - \bar{X})^2$
5	25	0	0
6	36	1	1
4	16	−1	1
8	64	3	9
2	4	−3	9
$G = \sum x = 25$	$\sum x^2 = 145$		$\sum (x - \bar{x})^2 = 20$

$$\text{Sum of Squares(SS)} = \sum (x - \bar{x})^2 = 20$$

where $\bar{x} = \frac{G}{N}$

By using the formula (2.4) the sum of squares in the above mentioned problem becomes

$$SS = \sum x^2 - \frac{G^2}{N} = 145 - \frac{25^2}{5} = 20$$

Thus, it can be seen that both the formulas (2.3) and (2.4) gives the same results in computing the sum of squares.

Because of its simplicity the formula (2.4) shall be used hereafter to compute various sum of squares in ANOVA.

ONE WAY ANALYSIS OF VARIANCE FOR INDEPENDENT MEASURES DESIGN

ANOVA is a group of statistical techniques used for analyzing different types of experimental designs. This technique is used for comparing means of three or more groups by comparing the sum of squares between groups with that of sum of squares within groups. The F-ratio is obtained by dividing the mean sum of squares between groups by the mean sum of squares within groups. As per the Snedecor, this F statistic follows F-distribution with (v_1, v_2) df, where v_1 is the degrees of freedom of between groups sum of square and v_2 is the degrees of freedom of within group sum of square. As per the central limit theorem, if groups are drawn from the same population, the variance between the group means should be lower than the variance within the groups. Thus, a higher value of F indicates that the samples have come from different populations. If in the design different levels of one factor are to be compared, the ANOVA is known as one-way ANOVA. In one-way ANOVA, it is essential that the subjects or units on which these different levels of one factor is applied should be homogeneous. Similarly if the effect of two or three factors on some criterion measure is investigated simultaneously, the ANOVA is said to be the two-way ANOVA or three-way ANOVA, respectively.

Assumptions

Following are the assumptions for one-way ANOVA in analyzing independent measures design:

a. The samples have been drawn from the population which is normally distributed.

b. Samples are independent to each other. In other words, subjects are different in each group.

c. The observations obtained on the subjects are independent to each other.

d. The populations from which the samples have been drawn have equal variances.

e. All the factors are additive in nature.

ILLUSTRATION I   25

Table 2.2   **Pull-ups Scores in Different Strength Training Groups**

Strength Training		
Low	Medium	High
5	8	9
3	6	8
4	5	7
5	4	8
2	3	6

The procedure involved in one-way ANOVA for independent measures design shall be discussed with the help of the following illustration.

ILLUSTRATION I

In an experiment it is required to compare the effect of three levels of strength training programs with different intensities (low, medium, and high) on the pull-ups performance obtained on college athletes. Fifteen subjects have been randomly assigned to three treatment groups. After training, their pull-ups performance has been recorded which are shown in Table 2.2. Let us see how to solve this one-way ANOVA for the independent measures design to find the most effective training program. The hypothesis shall be tested at the significance level of .05.

Solution

Here n = number of subjects in each treatment = 5

r = number of treatments=3

N = total number of subjects in the sample = nr = 15

Null hypothesis which is required to be tested is as follows:

$$H_0 : \mu_{low} = \mu_{medium} = \mu_{high}$$

against the alternative hypothesis that at least one of the group means differ.

To test the above mentioned null hypothesis, it is required to compute F ratio by computing sum of squares between groups and within groups. Since in this design there is only one source of variance (different levels of strength training), the total variation is distributed between groups and within groups. The sum of squares within groups is also known as sum of squares due to error. Thus, the one-way ANOVA for independent measures design, the total sum of squares can be partitioned as follows:

$$\text{Total SS} = \text{SS}_{Bet} + \text{SS}_{Error} \qquad (2.5)$$

where

Total SS	=	Variation of all the scores around combined mean value.
SS_{Bet}	=	Variation of group means around the combined mean
SS_{Within}	=	Combined variation of scores within each group taken around respective group means

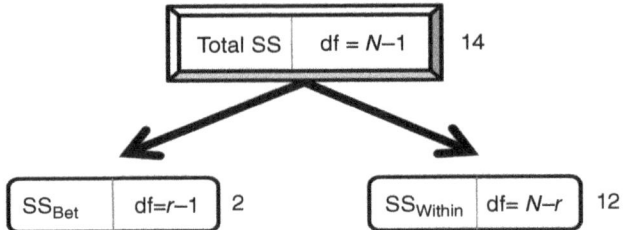

Figure 2.1   Scheme of distributing sum of squares and degrees of freedom

Partitioning of Total Variation in the Design

The total sum of squares and the degrees of freedom for one-way ANOVA as explained in Illustration I is distributed as per scheme shown in Figure 2.1.

Since computation of F depends upon the values of sum of squares between groups and within groups, the following computation shall be done to find these sums of squares.

Computation

$$\text{Correction factor (CF)} = \frac{G^2}{N} = \frac{83^2}{15} = 459.27$$

$$\text{Raw Sum of Squares (RSS)} = \sum_i \sum_j Y_{ij}^2$$

$$= (5^2 + 3^2 + \ldots 2^2) + (8^2 + 6^2 + \ldots + 3^2) + (9^2 + 8^2 + \ldots 6^2)$$

$$= 79 + 150 + 294$$

where Y_{ij} is the observation of the jth subject given the ith level of training, $i = 1,2,3$; $j = 1, 2, 3, 4, 5$.

$$\text{Total Sum of Square (TSS)} = \text{RSS} - \text{CF} = \sum_i \sum_j Y_{ij}^2 - \frac{G^2}{N}$$

$$= 523 - 459.27 = 63.73$$

$$\text{SS between treatment groups (SS}_b) = \sum_i \frac{R_i^2}{n_i} - \frac{G^2}{N}$$

$$= \frac{19^2}{5} + \frac{26^2}{5} + \frac{38^2}{5} - 459.27$$

$$= 72.2 + 135.2 + 288.8 - 459.27 = 36.93$$

where R_i is the ith group total and n_i the number of observation in the ith group, $i=1, 2, 3$.

ILLUSTRATION I 27

$$\text{SS within groups } (SS_w) = TSS - SS_b$$
$$= 63.73 - 36.93 = 26.8$$

Explanation

The following explanation related to different components worked out in this illustration shall help the readers to understand the procedure in a better way.

Partitioning of SS and Degrees of Freedom If you look into the pull-ups scores in Table 2.2, all the scores are not alike. One of the obvious reasons for the difference may be the training program. Since these scores have resulted due to different intensities of strength training program, the training may be one of the causes of variation. Thus, the first cause of variation is the training group. But within each group also all the scores are not same. This may be termed as individual variation. The individual variation arises in each group due to varied background of the subjects such as variation in their physical, physiological, and nutritional parameters. Since these individual variations cannot be assigned due to any particular cause, the pooled individual variations of all the groups is termed as experimental error. The experimental error is also known as within group error because it is sum of the individual variations in each group. Thus, the total variation in the scores is partitioned into two components; variation due to between groups and variation due to within groups.

Similarly total degrees of freedom, that is, $N - 1(= 14)$ has been partitioned into $r - 1(= 2)$ df for treatment groups and $N - r \,(= 12)$ df for within groups or due to error.

Computation Various terms computed above in the example will now be explained for understanding the procedure of computing F value.

Correction Factor The correction factor has been obtained by squaring the grand total of all the scores, G and dividing it by the total number of scores N. Thus, the correction factor has been obtained as 459.27.

Raw Sum of Squares The raw sum of squares is obtained by squaring all the scores in the study and adding them together. In this illustration there are $N \,(= 15)$ scores; hence all the 15 scores have been squared and added to get the raw sum of squares as 523.

Total Sum of Squares The total sum of squares indicates the variation among all the scores (irrespective of groups) around grand mean. The formula for computing SS as shown in (2.4) has been used in computing this total sum of squares.

$$SS = \sum x^2 - \frac{G^2}{N}$$

Table 2.3   Computation in One-Way ANOVA

	Strength Training			
	Low	Medium	High	
	5	8	9	
	3	6	8	
	4	5	7	
	5	4	8	
	2	3	6	
Group total	$R_1 = 19$	$R_2 = 26$	$R_3 = 38$	$G = R_1 + R_2 + R_3 = 83$
Group mean	3.8	5.2	7.6	

here x represents the score, N, the total number of scores, and G, the grand total of all the 15 scores. Thus, the total SS has been obtained by solving the following expression which is equal to 63.73.

$$= \underbrace{(5^2 + 3^2 + \ldots 2^2) + (8^2 + 6^2 + \ldots + 3^2) + (9^2 + 8^2 + \ldots 6^2)}_{\text{All fifteen scores in the study}} - \frac{83^2}{15} = 63.73$$

Sum of Squares between Groups   The SS between groups is the variation among group means around grand mean G/N. This term has been computed by using the following steps:

   i. Square the sum of each group, divide them by the number of scores in that group, and add them that is:

$$\frac{19^2}{5} + \frac{26^2}{5} + \frac{38^2}{5}, \text{ and}$$

  ii. Subtract the correction factor from the sum obtained in the step (i) to get the value of SS between groups which has been obtained as 36.93.

Sum of Squares within Groups   The SS within group is 26.8. It has been obtained by subtracting the sum of squares between groups from the total sum of squares, that is, 63.73−36.93. However, sum of squares can directly be calculated from the data in Table 2.3. The sum of squares within groups is the pooled sum of squares within each group. In other words, it is the sum of the variation within each of the three groups in this illustration. Let us see how to compute it. By using the formula (2.4) the sum of squares in each group can be computed as follows:

$$\text{SS within Low treatment group} = (5^2 + 3^2 + 4^2 + 5^2 + 2^2) - \frac{19^2}{5}$$

$$= 79 - 72.2 = 6.8$$

$$\text{SS within Medium treatment group} = (8^2 + 6^2 + 5^2 + 4^2 + 3^2) - \frac{26^2}{5}$$

$$= 150 - 135.2 = 14.8$$

ILLUSTRATION I   29

Table 2.4   ANOVA Table for the Data on Pull-ups

Source of Variation	df	SS	MSS	F-Value	Tab.F
Bet. groups	$r - 1 = 2$	36.93	$\frac{36.93}{2} = 18.47$	8.28^a	$F_{.05}(2,12) = 3.88$
Within groups	$N - r = 12$	26.8	$\frac{26.8}{12} = 2.23$		
Total	$N - 1 = 14$	63.73			

[a]Significant at 0.05 level

$$\text{SS within High treatment group} = (9^2 + 8^2 + 7^2 + 8^2 + 6^2) - \frac{38^2}{5}$$

$$= 294 - 288.8 = 5.2$$

Thus,

$$\text{SS within group} = 6.8 + 14.8 + 5.2 = 26.8$$

This value is exactly same as what has been obtained by the subtraction method.

ANOVA Table   Finally these sums of squares have been placed in the Table 2.4. This table is known as ANOVA table. The mean sum of squares between groups has been obtained by dividing SS between groups by the corresponding degrees of freedom $(r - 1)$. Similarly mean sum of squares within groups has been obtained by dividing the value of SS within groups by the corresponding degrees of freedom $(N - r)$. Finally, the F value has been computed by dividing mean sum of squares between groups by the mean sum of squares within groups.

Results

The tabulated value of F at .05 level of significance with df(2,12) obtained from the Table A.2 in the Appendix is 3.88. Since calculated value of F (= 8.28) is greater than the tabulated F, the null hypothesis is rejected and it may be concluded that the F value is significant. Hence we conclude that there is a significant difference among three levels of strength training. Since null hypothesis has been rejected, a post hoc test needs to be done to compare the means of different groups.

Post-Hoc Analysis

In order to find as to which group's pull-ups score is the best, a post hoc test shall be used. Since the sample size is same the Tukey's, HSD test shall be used for computing critical difference. Tukey's HSD test is also referred as Tukey's test. This test is used in post hoc analysis to test the significance of difference between two group means. Tukey's statistic compares all possible pairs of means and is based on a Studentized range distribution (q).

The critical difference in Tukey HSD test is given by

$$\text{CD} = q_{\alpha, r, N-r} \sqrt{\frac{\text{MSS}_w}{n}} \qquad (2.6)$$

where r is the number of groups, n is the number of scores in each group, and N is the total number of scores. The value of q at a particular significance level α and $(r, N - r)$ df can be obtained from the Table A.4 in the Appendix.

Here $\alpha = 0.05$, $r = 3$, $n = 5$, the critical difference (CD) shall be

$$CD = q_{0.05,3,12} \sqrt{\frac{2.23}{5}}$$

From the Table A.4 in the Appendix the value of $q_{0.05,3,12}$ is 3.77
Thus,

$$CD = 3.77 \times \sqrt{\frac{2.23}{5}} = 2.52$$

After arranging group means and critical difference in Table 2.5, group means can be compared in pairs for significant difference. The means of different groups have been written in descending order.

Looking to the Table 2.5 it is clear that the mean difference of pull-ups scores between high and low intensity groups is significant because it is higher than the critical difference. However, there is no significant difference between the mean pull-ups score in high and medium intensity groups as well as medium and low intensity groups. By looking to the mean values, it may be inferred that the pull-ups performance in the high intensity group is the highest whereas least in low intensity group.

It may therefore be inferred that the high intensity strength training program is the best in improving the pull-ups performance of the college athletes. The above results can be shown graphically in Table 2.6.

Table 2.5   Post-Hoc Comparison of Means Using Tukey Test

Means Pull-ups in Different Groups			Mean Diff.	CD at 5% Level
High	Medium	Low		
7.6	5.2		2.4	2.52
7.6		3.8	3.8[a]	2.52
	5.2	3.8	1.4	2.52

[a]Significant at 0.05 level.

Table 2.6   Mean Pull-Ups in Different Strength Training Groups

High	Medium	Low
7.6	5.2	3.8

"‿" represents no significant difference between the means at .05 level.

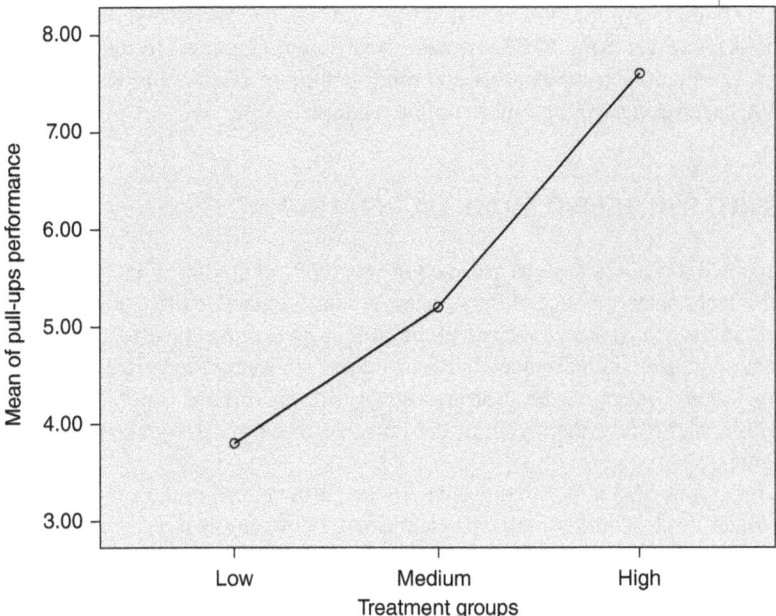

Figure 2.2   Means plot of the pull-ups performance in three strength training groups

Means Plot

The mean pull-ups score of all the three treatment groups can be plotted as shown in the Figure 2.2 to compare the effect of three different levels of strength training program on pull-ups performance.

The means plot shown in Figure 2.2 clearly indicates that the pull-ups performance of the high intensity groups is the best in comparison to that of low intensity program.

REPEATED MEASURES DESIGN

Repeated measures design is an extension of paired *t* test. In this design same subjects are tested repeatedly under all the treatment conditions; hence the name repeated measures. Repeated measures design is solved by using one-way ANOVA but for related, not independent groups. A repeated measures ANOVA is also referred to as a within-subjects ANOVA and may be written as rANOVA. In using this design researcher's interest is to investigate overall differences among related group means. For instance, if the effect of intervention (active, passive, and meditative) on the recovery is to be seen among the soccer players after the match, an experiment may be planned by using the repeated measures design. In this design all the subjects need to be repeatedly observed for their recovery response in all three different treatment conditions. Here all three groups are related because the same subjects are tested under all the three treatment conditions. The researcher's interest is to compare overall differences among these three related group means of recovery response.

In this book many complex repeated measures designs have been discussed along with their solutions using SPSS software in different chapters. To understand these designs, the manual computations and other details of one-way repeated measures ANOVA have been shown by means of an example.

WHEN TO USE REPEATED MEASURES ANOVA

The repeated measures designs are used in two types of studies. First type of studies are those where performance of the subjects are investigated in three or more different time durations. For instance the researcher might investigate the effect of an aerobic program on muscular endurance after four weeks, six weeks, and eight weeks. These three variations of the time duration are known as levels of Time factor. If five subjects are used in such study, the layout of the repeated measures design can be shown by the Figure 2.3.

Second types of studies are those where the performance of the subjects are compared under three or more treatment conditions. For example a researcher may wish to investigate the effect of sleep deprivation on EEG. He may choose three treatment conditions such as 24 hours, 30 hours, and 36 hours before the EEG examination. All the subjects may be tested for their EEG reading in all the three treatment conditions. In such studies the researcher must ensure that there is no carryover or learning effect due to the treatment conditions. To wipe off this effect, sufficient time should be given between testing a subject under any two treatment conditions. Another issue in such studies is that of the order effect. To done away the order effect counterbalancing is used in the design. This is done by dividing the sample into three groups if there are three treatment conditions. The first group will undergo the first treatment thereafter second and then the third treatment, whereas the second group may undergo into the second treatment first and then the first and subsequently the third treatment in sequence. Similarly in third group the order of treatment may be the third treatment first then the second and thereafter the first. If in the sleep deprivation example sample consists of six subjects then the layout of the one-way repeated measures design can be shown by the Figure 2.4. Here, in the first phase of testing the subjects S1 and S2 will be deprived of sleep for 24 hours, the subjects S3 and S4 for 30 hours and the subjects S5 and S6 for 36 hours. Similarly the testing protocol for the subjects during second and third phase of testing has been shown in the Figure 2.4.

Assumptions

In using repeated measures design for testing the hypothesis of interest certain assumptions are required to be fulfilled. In case the assumptions are violated the level of significance is inflated and the power of test reduces. In other words, the probability of wrongly rejecting the null hypothesis increases. The assumptions that are required for the repeated measures design are as follows:

1. The independent variable needs to be categorical (nominal or ordinal) and dependent variable to be metric (interval or ratio).

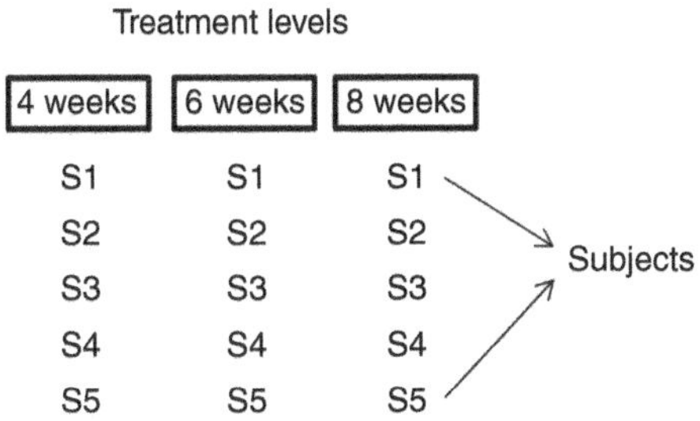

Treatment levels

Figure 2.3   Layout of the one-way repeated measures design having levels of the factor as time point.

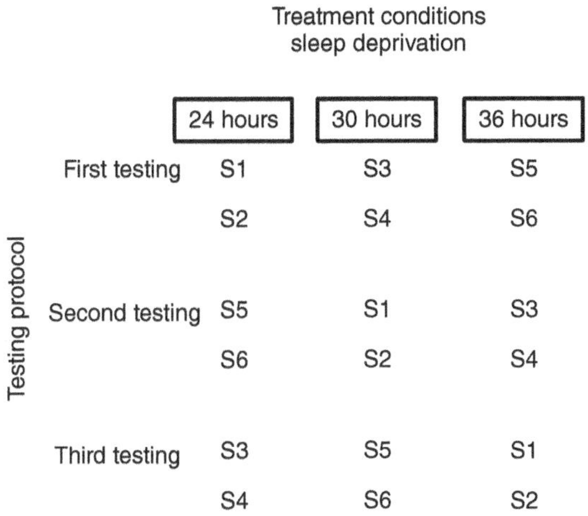

Treatment conditions
sleep deprivation

Figure 2.4   Layout of the one-way repeated measures design having three treatment conditions

2. There should be no outliers.
3. The difference scores in the dependent variable between the two or more related groups should approximately be normally distributed.
4. The assumption of sphericity must be satisfied. In other words, no sphericity should exist in the data. The sphericity assumption states that the variances of the differences between all combinations of related groups must be equal. In other words, correlations among the repeated measures are all equal.

SOLVING REPEATED MEASURES DESIGN WITH ONE-WAY ANOVA

The repeated measures design is also analyzed by using the ANOVA technique but for related groups. This section will help the reader to understand the computation used in repeated measures design by knowing the scheme of distributing total sum of squares and degrees of freedom. After going through the following sections readers will understand the difference between independent measures and repeated measures designs. The whole process of solving the one-way repeated measures design shall be discussed by means of an illustration. In solving the one-way repeated measures ANOVA one should use the following steps.

a. Hypothesis construction
b. Layout of the design
c. One-way repeated measures ANOVA model
d. Computation in one-way repeated measures ANOVA
e. Testing sphericity assumption
f. Testing significance of treatment effect
g. Writing results

ILLUSTRATION II

A researcher wanted to investigate the effect of music (classical, western, and Instrumental) on the mood of the subjects during dinner. A study was planned in which five subjects participated. In order to have control in the experiment similar menu at the same venue was kept in the experiment. The subjects were tested for their mood after each dinner session with different music and the scores so obtained were recorded which are shown in the Table 2.7. The counterbalancing was done while implementing the treatments so as to done away the effect of any systematic variance in the study. Let us apply one-way repeated measures ANOVA to find as to which type of music enhances the mood status during dinner. Higher the mood score more pleasurable state the subject is in. We shall test the hypothesis at .05 level of significance.

To solve this repeated measures design where all the five subjects were tested under each treatment condition, the following steps shall be performed.

Hypothesis Construction

The following null hypothesis is required to be tested

$$H_0 : \mu_{\text{Classical}} = \mu_{\text{Western}} = \mu_{\text{Instrumental}}$$

against alternative hypothesis that at least one of the group means differ.

In a repeated measures design the variability due to subject can be isolated from the treatment and error terms resulting reduction in the error sum of squares. This makes this design more efficient than the independent measures design. Here the TSS is split into SS between groups (treatment) and SS within groups. Further, SS within groups is divided into SS due to subjects and SS due to error. Thus, in one-way ANOVA with repeated measures, the model becomes

$$\text{Total } SS = SS_{\text{Treatment}} + SS_{\text{Within\_Treatmet}}$$
$$= SS_{\text{Treatment}} + SS_{\text{Subject}} + SS_{\text{Error}} \qquad (2.7)$$

ILLUSTRATION II					35

Table 2.7   Mood Score of the Subjects After Each Dinner Session

Subjects	Dinner Session with Music		
	Classical	Western	Instrumental
S1	4	6	2
S2	5	8	3
S3	6	7	4
S4	3	8	2
S5	4	9	3

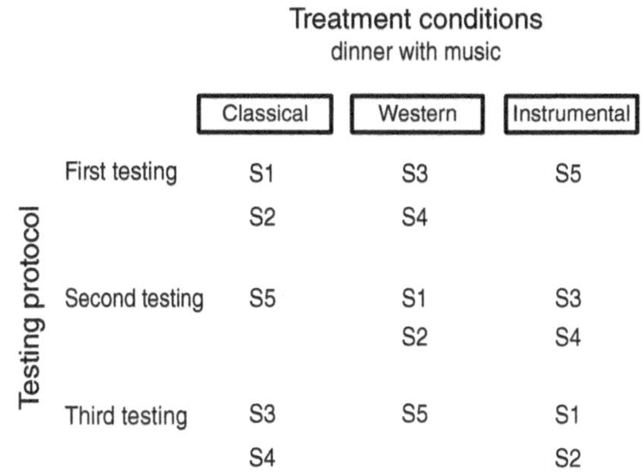

Treatment conditions
dinner with music

	Classical	Western	Instrumental
First testing	S1	S3	S5
	S2	S4	
Second testing	S5	S1	S3
		S2	S4
Third testing	S3	S5	S1
	S4		S2

Testing protocol

Figure 2.5 Layout of the one-way repeated measures design having three treatment conditions

where

Total SS	=	Variation of all the scores around combined mean value.
$SS_{Between}$	=	Variation of treatment group means around the combined mean
$SS_{Subject}$	=	Variation of within subjects scores around its mean
SS_{Error}	=	Variation of within treatment groups excluding individual variations

Layout Design

Since this is a repeated measures design where all the five subjects have been tested, it is necessary to use the counterbalancing in administering the treatments on the subjects. The design shown in Figure 2.5 shall explain the way treatment should be implemented on the subjects.

In this design on the first testing day the subjects S1 and S2 have been tested for their mood after dinner with classical music, the subjects S3 and S4 with western

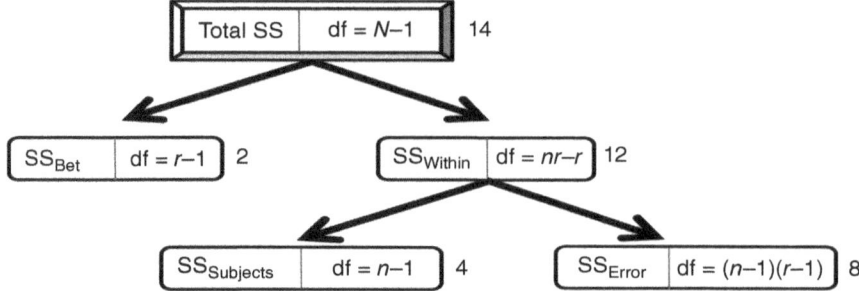

Figure 2.6 Scheme of distributing total sum of squares and degrees of freedom in one-way repeated measures design

music, whereas the subject S5 with instrumental music. Similarly, on second and third testing day subjects have received different treatments as shown in the Figure 2.5. By changing the order of implementing the treatment on the subjects the effect of any systematic variance in the study is removed. This is also known as counterbalancing of treatment implementation.

One-Way Repeated Measures ANOVA Model

Unlike one-way ANOVA for independent measures, in repeated measures design subjects serve their own control and therefore part of the experimental variability is explained by the subjects. Due to this, error variance gets reduced resulting increase in the F value. Thus, even small effect in the criterion variable due to the change in the independent variable can be detected. In this design splitting the total sum of squares into different components along with their degrees of freedom can be shown by the Figure 2.6.

From the Figure 2.6 you can see that the sum of squares within groups has been divided further into sum of squares due to subjects and sum of squares due to experimental error. In this design the actual error is given by SS_{Error} instead of SS_{Within} in independent measures design. The experimental error has been reduced in this design because part of the variability within groups is explained by the subjects. Reduction of this error variance makes this design more robust in comparison to independent measures design.

In repeated measures design the total degrees of freedom, that is, $N - 1(= 14)$ has been partitioned into $r - 1(= 2)$ df due to variation in treatment groups and $N - r(= 12)$ df due to variation within treatment groups. Since SS within treatment is further divided into SS due to subjects and SS due to error, the degrees of freedom for SS within treatment $nr - r(= 12)$ is further split into $n - 1(= 4)$ df due to subjects and $(n - 1)(r - 1)[= 8]$ df due to error.

Computation in Repeated Measures Design with One-Way ANOVA

Here,
   Number of treatments $= r = 3$
   Number of subjects $= n = 5$

ILLUSTRATION II   37

$$\text{Total number of scores} = N = nr = 15$$

$$\text{Correction factor (CF)} = \frac{G^2}{N} = \frac{74^2}{15} = 365.07$$

$$\text{Raw Sum of Squares (RSS)} = \sum_i \sum_j Y_{ij}^2$$

$$= (4^2 + 6^2 + 2^2) + (5^2 + 8^2 + 3^2)$$

$$+ \dots + (4^2 + 9^2 + 3^2)$$

$$= 438$$

where ($i = 1, 2, 3$; i represents the treatment group) ($j = 1, 2, \dots 5$; j represents subject)

$$\text{Total Sum of Squares (TSS)} = \text{RSS} - \text{CF} = \sum_i \sum_j Y_{ij}^2 - \frac{G^2}{N}$$

$$= 438 - 365.07 = 72.93$$

$$\text{SS between Treatments (SS}_{\text{Treatments}}) = \sum_j \frac{T_j^2}{n} - \frac{G^2}{N}$$

$$= \frac{22^2}{5} + \frac{38^2}{5} + \frac{14^2}{5} - 365.07$$

$$= 96.8 + 288.8 + 39.2 - 365.07 = 59.73$$

$$\text{SS between Subjects (SS}_{\text{Subjects}}) = \sum_i \frac{S_i^2}{r} - \frac{G^2}{N}$$

$$= \frac{12^2}{3} + \frac{16^2}{3} + \frac{17^2}{3} + \frac{13^2}{3} + \frac{16^2}{3} - 365.07$$

$$= 371.33 - 365.07$$

$$= 6.26$$

$$\text{SS due to Error (SS}_{\text{Error}}) = \text{TSS} - \text{SS}_{\text{Bet}} - \text{SS}_{\text{Subjects}}$$

$$= 72.93 - 59.73 - 6.26$$

$$= 6.94$$

Explanation

The following explanation related to different components worked out in this repeated measures design shall help the readers to understand the analysis in a better way.

Computation Various terms computed above in this example will now be explained for understanding the procedure of computing F value in this illustration of repeated measures design.

Correction Factor   The correction factor has been obtained by squaring the grand total of all the scores $G (= 74)$ and dividing it by the total number of scores $N (= 15)$. Thus, the correction factor has been obtained as 365.07.

Raw Sum of Squares   The raw sum of squares is obtained by squaring all the scores in the study and adding them together. Since in this illustration $N (= 15)$ scores are there, all the 15 scores have been squared and added which gives the value of raw sum of squares as 438.

Total Sum of Squares   The total sum of squares indicates the variation among all the scores irrespective of groups around the grand mean. The formula for computing SS as shown in (2.4) has been used in computing this total sum of squares.

$$\text{TSS} = \sum x^2 - \frac{G^2}{N}$$

here x represents the scores in the study, N is the total number of scores, and G represents the grand total of all the 15 scores. Thus, the total SS has been obtained by solving the following expression which is equal to 72.93.

$$\text{Total SS} = \underbrace{(4^2+5^2+......4^2) + (6^2+8^2+......+9^2) + (2^2+3^2+......3^2)}_{\text{All fifteen scores in the study}} - \frac{74^2}{15} = 72.93$$

Sum of Squares between Treatments   The SS between treatments is the variation among treatment group means around grand mean G/N. This term has been computed by using the following steps:

i. Square the sum of each treatment group, divide them by the number of scores in that group, and add them, that is,

$$\frac{22^2}{5} + \frac{38^2}{5} + \frac{14^2}{5}, \text{ and}$$

ii. Subtract the correction factor from the sum obtained in the step (i) above, to get the value of SS between treatment groups which has been obtained as 59.73.

Sum of Squares between Subjects   The SS between subjects refers to the variation among subject's average scores (in all the treatment conditions) around grand mean G/N. You can obtain this term by following the below mentioned steps:

i. Square the sum of each subject's scores in all the treatment, divide them by the number of treatments in that group, and add them, that is,

$$\frac{12^2}{5} + \frac{16^2}{5} + \frac{17^2}{5} + \frac{13^2}{5} + \frac{16^2}{5}, \text{ and}$$

ii. Subtract the correction factor from the sum obtained in the step (i) to get the value of SS between subjects which has been obtained as 6.26.

ILLUSTRATION II   39

Table 2.8   Computation in One-Way Repeated ANOVA

Subject	Classical	Western	Instrumental	Total (S_i)
	Workout Sessions with Music			
S1	4	6	2	12
S2	5	8	3	16
S3	6	7	4	17
S4	3	8	2	13
S5	4	9	3	16
Group total	$T_1 = 22$	$T_2 = 38$	$T_3 = 14$	$G = R_1 + R_2 + R_3 = 74$
Group mean	4.4	7.6	2.8	

Sum of Squares Due to Error   The sum of squares due to error is the residual error in the experiment. This has been obtained by subtracting the SS between treatment and SS between subjects from the total sum of squares. The sum of squares due to error has been obtained as 6.94 by following this subtraction method.

However, this sum of squares can also be computed directly from the data in Table 2.8. Since the error SS is the residual error which is obtained by subtracting the SS between treatment and SS between subjects from the total SS.

Thus,

SS due to error

$$= \left[\left(4^2 + 5^2 + \ldots 4^2\right) + \left(6^2 + 8^2 + \ldots + 9^2\right) + \left(2^2 + 3^2 + \ldots 3^2\right) - \frac{74^2}{15} \right]$$

$$- SS_{\text{Treatment}} - SS_{\text{Subjects}}$$

$$= 72.93 - 59.73 - 6.26 = 6.94$$

This value is exactly same as what has been obtained by the subtraction method.

ANOVA Table   All the sums of squares so calculated have been placed in the Table 2.9. This table is known as ANOVA table for repeated measures. In repeated measures design we are only concerned about the F value for treatment and not for subject. Hence F value for treatment has been computed. Since the same subjects are tested in all the three treatment conditions, the significance of F will not be tested in the usual manner the way it was done in case of independent measures design. Since the subjects in all the treatments are not distinct, it is required to test the assumption of sphericity first before drawing any conclusion about the F value.

Testing Sphericity Assumption

One of the main assumptions in using repeated measures design is the sphericity. It is assumed that there is no sphericity present in the data generated in the study. In case sphericity assumption is violated some correction is required to be made in testing desired hypothesis. If sphericity assumption violates the power of the test decreases. In other words the level of significance inflates. More specifically, probability of wrongly rejecting the null hypothesis increases in case of violation of this assumption. The sphericity refers to a situation where the variances of the differences between all possible pairs of groups are equal. If sphericity assumption holds in this illustration then the variance of the difference scores obtained in each pair i.e. classical-western, classical-instrumental, and western-instrumental should be same. But the variance of these three difference groups are 2.56. 0.24 and 1.36 as shown in Table 2.10. The next question is whether these variances can be considered to be equal or there is a significant difference among them.

The null hypothesis of no difference among these three variances may be tested by using the Mauchly's test of sphericity. The SPSS provides this test while solving the repeated measures design. Since Mauchly's statistic follows chi-square distribution, chi-square statistic is used for testing the significance of sphericity. For this data set the value of chi-square (χ^2) is 8.129 as shown in Table 2.11. This value of χ^2 is significant as its associated p value is 0.017 which is less than 0.05. The detailed procedure of using SPSS for one-way ANOVA for repeated measures design has been shown in Chapter 4.

If the Mauchly's test is significant, the sphericity assumption is violated. Thus, if the Mauchly's test is not significant ($p > 0.05$), the significance of F is tested by assuming the sphericity and without doing any correction in the degrees of freedom of the treatment and error terms. In this case sphericity assumption is violated and the correction in the degrees of freedom is required to be made before testing the significance of F value.

Table 2.9   ANOVA Table for the Repeated Measures on the Data on Mood

Source of Variation	df	SS	MSS	F-Value	Tab.F
Treatment	$r - 1 = 2$	59.73	$59.73/2 = 29.87$	34.33[a]	$F_{.05}(2,8) = 4.46$
Subjects	$n - 1 = 4$	6.26	$6.26/4 = 1.57$		
Error	$(r - 1)(n - 1) = 8$	6.94	$6.94/8 = 0.87$		
Total	$N - 1 = 14$	72.93			

[a]Significant at 0.05 level.

Table 2.10   Computation for Sphericity

Classical	Western	Instrumental	Classical-Western	Classical-Instrumental	Western-Instrumental
4	6	2	−2	2	4
5	8	3	−3	2	5
6	7	4	−1	2	3
3	8	2	−5	1	6
4	9	3	−5	1	6
			$\sigma^2 = 2.56$	$\sigma^2 = 0.24$	$\sigma^2 = 1.36$

ILLUSTRATION II   41

The next question is to know as to how much sphericity violation has been made. The severity of sphericity violation is measured by a value known as epsilon (ε) as computed in Table 2.11. If the value of epsilon (ε) is 1, the sphericity is not violated, whereas lesser the value of epsilon (ε) greater is the violation of sphericity assumption. If sphericity assumption violates level of significance inflates and therefore to compensate this error correction in the degrees of freedom for the Treatment and Error terms is made which provides the corrected p value for testing the significance of F. The readers should note that due to change in the degrees of freedom, there won't be any change in the computed value of F but only significance value (p) would change.

To modify the degrees of freedom for treatment and error terms, three different corrections have been suggested, namely Greenhouse–Geisser, Huynh–Feldt, and Lower bound. Different values of epsilon (ε) are provided in these methods.

The lower bound method is considered to be the most conservative while the Greenhouse–Geisser is considered to be the more conservative and the Huynh–Feldt is the least conservative. The lower bound approach provides the greatest possible violation of sphericity, hence never used by the researcher. If the value of epsilon (ε) is greater than 0.75, the Greenhouse–Geisser correction tends to underestimate epsilon (ε), whereas the Huynh–Feldt correction tends to overestimate it. It is generally recommended to use the Greenhouse–Geisser correction, in case the estimated epsilon (ε) is less than 0.75 and Huynh–Feldt correction if it is more than 0.75.

Correcting for Degrees of Freedom   Let us see how the degrees of freedom for treatment and errors are corrected by using these corrections. These corrections are made by multiplying the epsilon (ε) value to the degrees of freedom of treatment and error terms.

Correction by Using Lower Bound   For lower bound the value of epsilon (ε) in Table 2.11 is 0.5, hence the new degrees of freedom for treatment and error would be as follows:

$$\text{Corrected degrees of freedom for Treatment} = \varepsilon \times (r - 1) = 0.5 \times (3 - 1) = 1$$

$$\text{Corrected degrees of freedom for Error} = \varepsilon \times (r - 1)(n - 1)$$

$$= 0.5 \times (3 - 1)(5 - 1) = 4$$

Table 2.11   Mauchly's Test of Sphericity

Measure: Mood							
Within Subjects Effect	Mauchly's W	Approx. Chi-Square	df	Sig.	Epsilon[a]		
					Greenhouse–Geisser	Huynh–Feldt	Lower Bound
Treatment	0.067	8.129	2	0.017	0.517	0.535	0.500

[a]May be used to adjust the degrees of freedom for the averaged tests of significance. Corrected tests are displayed in the tests of Within-Subjects Effects table.

Table 2.12   Tests of Within-Subjects Effect

		Measure: Mood					
Source		Type III SS	df	Mean SS	F	Sig.	Partial Eta Squared
Treatment	Sphericity assumed	59.733	2	29.867	34.462	0.000	0.896
	Greenhouse–Geisser	59.733	1.03	57.745	34.462	0.004[a]	0.896
	Huynh–Feldt	59.733	1.07	55.843	34.462	0.003	0.896
	Lower-bound	59.733	1.00	59.733	34.462	0.004	0.896
Error	Sphericity assumed	6.933	8	0.867			
	Greenhouse–Geisser	6.933	4.14	1.676			
	Huynh–Feldt	6.933	4.28	1.620			
	Lower-bound	6.933	4.00	1.733			

[a]Significant at 0.05 level.

where r is number of repeated measures and n is the number of subjects. In this illustration r is 3 and n is 5. Thus, in using the Lower bound correction, the critical value of F needs to be seen at (1,4) df instead of F at (2,8) df as shown in the Table 2.12. This has increased the p-value (= 0.004) to compensate the violation of sphericity.

Correction by Using Greenhouse–Geisser   For Greenhouse–Geisser the value of epsilon (ε) in Table 2.11 is 0.517, hence the new degrees of freedom for treatment and error would be as follows:

$$\text{Corrected degrees of freedom for Treatment} = \varepsilon \times (r - 1)$$

$$= 0.517 \times (3 - 1) = 1.03$$

$$\text{Corrected degrees of freedom for Error} = \varepsilon \times (r - 1)(n - 1)$$

$$= 0.517 \times (3 - 1)(5 - 1)$$

$$= 4.14$$

Thus, in using the Greenhouse–Geisser correction, the critical value of F needs to be seen at (1.03, 4.14) df instead of F at (2,8) df as indicated in the Table 2.12. This has increased the p-value (= 0.004) to compensate the violation of sphericity assumption.

Correction by Using Huynh–Feldt   For Huynh–Feldt the value of epsilon (ε) in Table 2.11 is 0.535, hence the new degrees of freedom for treatment and error would be as follows:

$$\text{Corrected degrees of freedom for Treatment} = \varepsilon \times (r - 1)$$

$$= 0.535 \times (3 - 1) = 1.07$$

ILLUSTRATION II   43

Table 2.13 Pair-Wise Comparisons

Measure: Mood						
Treatment (I)	Treatment (J)	Mean Diff (I − J)	SE	Sig.[a]	95% CI for Difference[a]	
					Lower Bound	Upper Bound
Classical	Western	−3.200[b]	0.800	0.048	−6.369	−0.031
	Instrumental	1.600[b]	0.245	0.009	0.630	2.570
Western	Classical	3.200[b]	0.800	0.048	0.031	6.369
	Instrumental	4.800[b]	0.583	0.004	2.490	7.110
Instrumental	Classical	−1.600[b]	0.245	0.009	−2.570	−0.630
	Western	−4.800[b]	0.583	0.004	−7.110	−2.490

Based on estimated marginal means
[a] Adjustment for multiple comparisons: Bonferroni.
[b] The mean difference is significant at the 0.05 level.

$$\text{Corrected degrees of freedom for Error} = \varepsilon \times (r-1)(n-1)$$
$$= 0.535 \times (3-1)(5-1)$$
$$= 4.28$$

Thus, in using the Huynh–Feldt correction, the critical value of F needs to be seen at (1.07, 4.28) df instead of F at (2,8) df. This has increased the p-value ($= 0.003$) to compensate the sphericity violation.

Results

In this illustration Mauchly's test shows that the sphericity has been violated, $\chi^2 = 8.129$, $p = 0.017$, therefore the degrees of freedom for treatment and error were corrected by using the Greenhouse–Geisser estimates of sphericity ($\varepsilon = 0.517$). This correction was chosen because the value of epsilon (ε) was less than 0.75. The result in Table 2.12 shows that the F value for the treatment is significant because p value associated with F after applying the correction is 0.004. It may therefore be concluded that the workout with change in the background music had a significant impact on the moods of the subjects. In order to know as to which treatment condition is the best in mood changing the pair-wise comparison of means shall be done.

Pair-Wise Comparison of Means

In repeated measures design there is no post hoc test for comparing group means. Pair-wise comparison of means is done by using the t-test for related groups. Due to multiple comparisons the level of significance (α) inflates. To compensate this loss of power, either Bonferroni or Sidak test can be used but the Bonferroni test is generally preferred for effective correction. The detailed procedure for pair-wise comparison of means has been shown in Chapter 4. For the data in Illustration 2, the pair-wise comparison of means has been shown in Table 2.13 by using the Bonferroni correction.

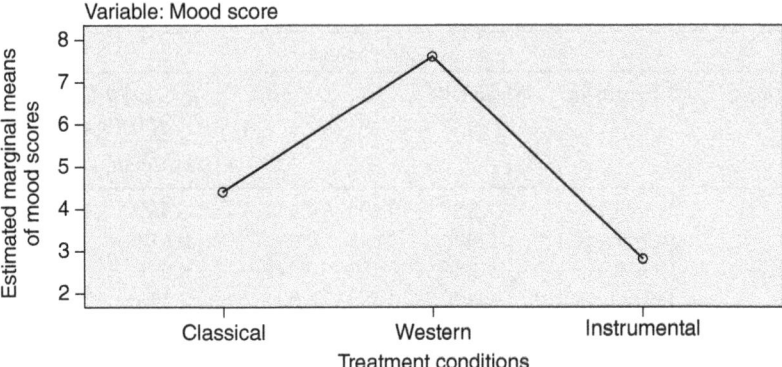

Figure 2.7   Means plot of the mood scores during workout with different types of music

It can be seen from the Table 2.13 that all the three group means differ significantly from each other. Thus, by looking to the mean scores on mood under all the three treatment conditions as shown in Table 2.8, it may be inferred that mood of the subjects is elevated maximum while doing the workout along with the western music and is lowest when workout is done with the instrumental music. This can be shown by the means plot in Figure 2.7.

BONFERRONI CORRECTION

The Bonferroni correction is made to adjust the p value when multiple comparisons are made simultaneously on a single set of data. Let us understand as to why such correction is required. In comparing two group means the t test and F test yields similar results because of the relationship between them, that is, $F = t^2$. Therefore, it is immaterial whether you apply one-way ANOVA or t test in comparing two group means. But if the number of groups is more than three, the ANOVA procedure is more efficient than using several t tests. In using multiple t-tests instead of one-way ANOVA, the Type I error inflates in testing the hypothesis. For instance in comparing four groups, six t tests need to be applied. If the hypothesis is tested at the significance level 0.05, the probability of committing Type I error in a single comparison is 0.05 and therefore the probability of not committing Type I error in a single comparison would be $1 - 0.05 = 0.95$.

$\therefore P$(Not committing Type I error in all the six comparisons)

$= 0.95 \times 0.95 \times 0.95 \times 0.95 \times 0.95 \times 0.95$

$= 0.95^6$

$\Rightarrow P$(Committing Type I error at least in any one of the six comparisons)

$= 1 - 0.95^6$

$= 1 - 0.7351 = 0.265$

Thus, we have seen that in applying six pair-wise comparisons, although the significance level was taken as 0.05, the actual level of significance has inflated to 0.265.

In repeated measures design post hoc tests cannot be applied; hence in a situation where the F value is significant one needs to know as to which treatment condition is the best. To do so one needs to have many pair-wise comparison of means. We have seen above that in case of making pair-wise comparisons in a situation where four treatment conditions are to be tested, the level of significance inflates from 0.05 to 0.265. Thus, probability of committing Type I error increases from 0.05 to 0.265 if six comparisons are made. In such situations Bonferroni correction is required to be made due to the loss of power in making multiple comparisons on the same set of date. In applying a Bonferroni correction, the critical value α is roughly divided by the number of comparisons required to be made. Thus, in the above example where six comparisons are tested for its significance, the level of significance should be taken as 0.008(0.05/6) instead of 0.05. In fact SPSS automatically apply this correction if you choose the option of Bonferroni correction for pair-wise comparisons.

EFFECT SIZE

The effect size is a useful measure for getting an idea about the effectiveness of a particular intervention in comparison to others. It gives you an idea about the worth of relationships between independent and dependent variables. For instance, an exercise program may be significant in enhancing the cardio respiratory endurance but whether this significance really encourages someone to follow that exercise regimen. This decision can be taken by looking to the value of the effect size. Now days it is customary to quote the effect size in the experimental findings. The SPSS software provides option to compute the effect size by means of partial eta square (η^2) while solving any design. The effect size is denoted by eta square or partial eta square and is computed by the following formula:

$$\eta^2_{partial} = \frac{SS_{treatments}}{SS_{treatmentss} + SS_{error}}$$

[If dependent variable is different treatment conditions]   (2.8)

or

$$\eta^2_{partial} = \frac{SS_{time}}{SS_{time} + SS_{error}}$$ [If dependent variable is different time points]

In the above mentioned illustration

$$\eta^2_{partial} = \frac{SS_{treatments}}{SS_{treatmentss} + SS_{error}}$$

$$= \frac{59.733}{59.73 + 6.933}$$

$$= 0.896$$

The values of eta square and partial eta square are same if there is only one independent variable. However, if the effect of more than one independent variable is investigated, effect size is reported as partial eta square. In fact the partial eta square is the variance explained by a given independent variable of the variance remaining after excluding variance explained by other independent variables in the design. As a thumb rule the value of η^2 is considered as small if it is 0.2, moderate if 0.5, and large if 0.8.

EXERCISE

2.1. What are the basic difference between independent measures design and repeated measures design? Explain by means of an example.

2.2. What do you mean by sum of squares and how it is computed? Following are the data on height of the students in different sports groups. Compute sum of squares between groups and within groups.

Height of the College Students		
Gymnastics	Swimming	Volleyball
174	180	185
170	178	184
172	176	182

2.3. Compare the strength and weekness of one-way independent measures design and one-way repeated measures design. Support your answer by means of distributing sums of squares and degrees of freedom.

2.4. Discuss the assumptions in independent measures design. What happens if these assumptions are not satisfied?

2.5. Discuss the situations where the repeated measures designs can be used. Explain by means of an example.

2.6. If effect of three environments needs to be compared on functional efficiency of nine subjects during daylong working in an organization in a repeated measures design, draw the layout of design and partitioning of variation.

2.7. How is it different in using pair-wise comparison of means in independent measures design and repeated measures design?

2.8. Discuss the assumptions in independent and repeated measures designs. How will you tackle if these assumptions are not met?

2.9. What do you mean by sphericity? Explain by taking an example. What happens if sphericity exists in the repeated measures design?

2.10. Discuss the procedure in solving a repeated measures design.

2.11. What is Bonferroni correction in repeated measures design?

2.12. What is Effect size? What is its significance in showing the results?

ASSIGNMENT

2.1. Compute sum of squares for the following set of scores:

$$X: \quad 3 \qquad 79 \qquad 56 \qquad 84 \qquad 63 \qquad 7$$

2.2. If in a repeated measures design four levels of within-subjects variable are compared at the significance level of 0.05, how much the level of significance would be inflated?

2.3. Following are the scores on motivation of the employees in three different departments of an industry. Apply one-way ANOVA to compare motivation level of employees in these departments and discuss your findings at the significance level 0.05. Also do the following:

a. Show the scheme of partitioning the total sum of squares into different components along with the degrees of freedom.

b. Name the independent and dependent variables in this study?

c. If average motivation level among these three groups differs what conclusion can you draw in terms of relationship between independent and dependent variables?

d. Check your findings either with Excel or SPSS

	Scores on Motivation		
SN	Production	Marketing	Human Resource
1	23	39	28
2	28	36	25
3	25	41	36
4	31	40	34
5	27	38	32
6	26	37	28
7	29	43	34
8	32	36	41
9	38	32	30
10	26	39	34

2.4. To see the impact of exercise equipments on sweat loss an exercise scientist organized an experiment using repeated measures design in which six subjects were asked to run on treadmill and perform an exercise on stepper and bicycle ergometer with same intensity for half an hour. The body weight of the subjects was measured before and after each exercise to know the sweat loss. The data so obtained are shown below. Solve this repeated measures design and discuss your findings at the significance level 0.05. Do the following also:

a. Describe the partitioning of total sum of squares into different components along with the degrees of freedom and explain how it is superior to independent measures design.
b. Name the independent and dependent variables in this study?
c. Check sphericity assumption manually.
d. Compute effect size.
e. Check your findings either with Excel or SPSS.

Sweat loss in grams			
Subject	Treadmill	Stepper	Bicycle Ergo Meter
1	150	160	200
2	148	165	190
3	160	165	155
4	145	170	210
5	135	160	190
6	145	165	195

BIBLIOGRAPHY

Anderson DR, Sweeney DJ, Williams TA. *Statistics for Business and Economics*. 6th ed. Minneapolis/St. Paul: West Pub. Co; 1996. p 452–453. ISBN: 0-314-06378-1.

Bland JM, Altman DG. Multiple significance tests: the Bonferroni method. BMJ 1995;310(6973):170.

Brand A, Bradley MT, Best LA, Stoica G. Multiple trials may yield exaggerated effect size estimates. J Gen Psychol 2011;138(1):1–11. DOI: 10.1080/00221309.2010.520360.

Cochran WG, Cox GM. *Experimental Designs*. 2nd ed. New York: Wiley; 1992. ISBN: 978-0-471-54567-5.

Cohen J. *Statistical Power Analysis for the Behavior Sciences*. 2nd ed. Routledge; 1988. ISBN: 978-0-8058-0283-2.

Conaway, M. Repeated Measures Design. 1999. Retrieved February 18, 2008, from http://biostat.mc.vanderbilt.edu/twiki/pub/Main/ClinStat/repmeas.PDF.

Cox DR (2006). *Principles of Statistical Inference*. Cambridge, New York: Cambridge University Press. ISBN: 978-0-521-68567-2.

Ellis PD. *The Essential Guide to Effect Sizes: An Introduction to Statistical Power, Meta-Analysis and the Interpretation of Research Results*. UK: Cambridge University Press; 2010.

Field AP. *Discovering Statistics Using SPSS*. Sage Publications; 2005.

Freedman DA. *Statistical Models: Theory and Practice*. Cambridge University Press; 2005. ISBN: 978-0-521-67105-7.

Gelman A. *"Variance, Analysis of"*. *The New Palgrave Dictionary of Economics*. 2nd ed. Basingstoke, Hampshire, New York: Palgrave Macmillan; 2008. ISBN: 978-0-333-78676-5.

Girden E. *ANOVA: Repeated Measures*. Newbury Park, CA: Sage; 1992.

Heiman GW. *Basic Statistics for the Behavioral Sciences*. 14 th ed. Boston, MA: Houghton Mifflin Company; 2003.

Hinkelmann, Klaus & Kempthorne, Oscar 2008. Design and Analysis of Experiments. I and II (Second ed.). John Wiley & Sons, Inc. ISBN 978-0-470-38551-7. 2008. Section 6.3

Kelley K, Preacher KJ. On effect size. Psychol Methods 2012;17(2):137–152. DOI: 10.1037/a0028086.

Lehmann EL. *Testing Statistical Hypotheses*. John Wiley & Sons; 1959.

Maxwell SE, Delaney HD. *Designing Experiments and Analyzing Data: A Model Comparison Perspective*. Belmont: Wadsworth; 1990.

Minke, A. Conducting Repeated Measures Analyses: Experimental Design Considerations. 1997. Retrieved February 18, 2008, from Ericae.net:http://ericae.net/ft/tamu/Rm.htm.

Moore DS, McCabe GP. *Introduction to the Practice of Statistics (4e)*. W H Freeman & Co.; 2003. ISBN: 0-7167-9657-0.

Nakagawa S, Cuthill IC. Effect size, confidence interval and statistical significance: a practical guide for biologists. Biol Rev Cambridge Philos Soc 2007;82(4):591–605. DOI: 10.1111/j.1469-185X.2007.00027.x. PMID 17944619.

Fisher RA. On the "probable error" of a coefficient of correlation deduced from a small sample. Metron 1921;1:3–32.

Pierce CA, Block RA, & Aguinis H. 2004. Cautionary note on reporting eta-squared values from multifactor ANOVA designs. Educational and Psychological Measurement, 64, 916–924.

Pollatsek A, Well AD. On the use of counterbalanced designs in cognitive research: a suggestion for a better and more powerful analysis. J Exp Psychol 1995;21:785–794.

Scheffé H. *The Analysis of Variance*. New York: Wiley; 1959.

Shaughnessy JJ. *Research Methods in Psychology*. New York: McGraw-Hill; 2006.

Snedecor GW, Cochran WG. 1967. Statistical Methods. Ames: Iowa State University (6): 253–256.

Stigler SM. *The History of Statistics : The Measurement of Uncertainty before 1900*. Cambridge, Mass: Belknap Press of Harvard University Press; 1986. ISBN: 0-674-40340-1.

Wilkinson L, APA task force on statistical inference. Statistical methods in psychology journals: Guidelines and explanations. Am Psychol 1999;54(8):594–604. DOI: 10.1037/0003-066X.54.8.594.

3

TESTING ASSUMPTIONS IN REPEATED MEASURES DESIGN USING SPSS

INTRODUCTION

Like independent measures design in repeated measures design also a researcher is interested in investigating the relationship between an independent and some dependent variable. In such studies focus is to find whether a treatment affects the performance or not. For instance, one may like to investigate the impact of advertisement on the sale of a product. A hypothesis may be tested that the advertisement increases the sale of a product. In such testing an experimenter may commit two kinds of error. Firstly, he may conclude that there is a relationship between advertisement and sales performance, whereas actually there is not. Secondly, he may conclude that there is no relationship between advertisement and sales performance, whereas actually it exists. The first type of erroneous decision is known as Type I error whereas the second is termed as Type II error. If assumptions of repeated measures designs are not fulfilled, then the type I error increases and the findings obtained in the study is not valid. In other words, there is an issue of conclusion validity if the assumptions are not met. In this chapter, the procedure of testing assumptions required for the repeated measures design shall be shown by means of SPSS. For those readers who are not acquainted with the SPSS, the next section shall help them to learn introductory steps. Understanding assumptions shall help readers to comprehend well the analysis part of solving different repeated measures designs discussed in different chapters. Further, a discussion shall be made on the determination of sample size towards the end of the chapter.

Repeated Measures Design for Empirical Researchers, First Edition. J. P. Verma.
© 2016 John Wiley & Sons, Inc. Published 2016 by John Wiley & Sons, Inc.

FIRST STEP IN USING SPSS

Before using SPSS, it needs to be activated on your computer. This can be done by using the command sequence: **Start → All Programs → IBM SPSS Statistics.** After following this command sequence you will get the screen as shown in Figure 3.1. The first step in using any analysis in SPSS is to prepare a data file. Check the option 'Type in data' on the screen shown in Figure 3.1 if you are entering the data for the first time. But if the data file has already been created then use the option 'Open an existing data source' for using the already saved file. Click on **OK** to get the screen as shown in Figure 3.2.

The screen shown in Figure 3.2 will facilitate you to create a new data file for the analysis. The data file in SPSS is created in two steps. First, all the variables in the study are defined by clicking on the **Variable View** option in the screen.

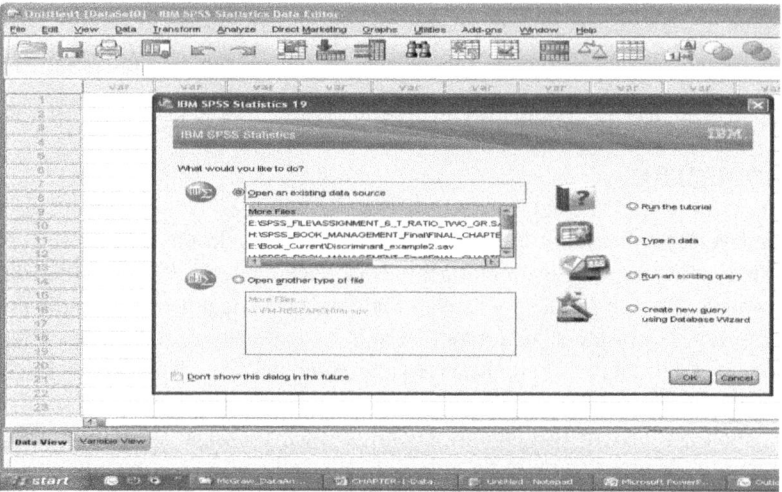

Figure 3.1   Screen showing option for creating/opening file

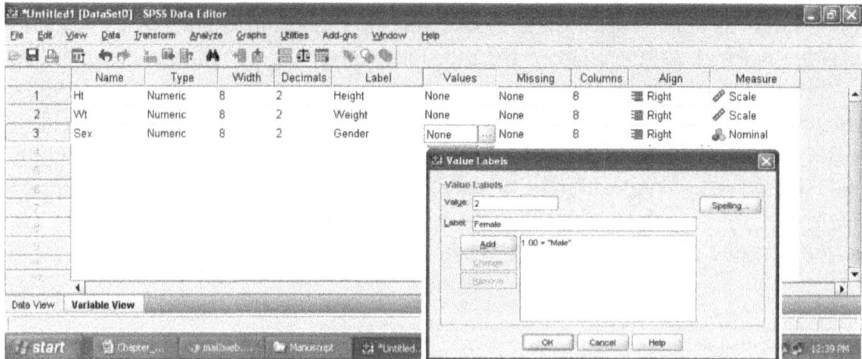

Figure 3.2   Screen showing option for defining variables and coding

In first column under the heading **Name** define short name of the variable. The variable name should essentially start with an alphabet and may include underscore and numerals in between, without any gap. There should be no space between any two characters of the variable name. If the name consists of two words, they must be joined by using the underscore. Further, variable name should not be started with numerals or any special character.

Under the column heading **Label**, full name of variable is defined. One can take advantage of this facility to write the expanded name of variable the way one feels like.

Under the column heading **Values** coding of the variable may be defined by double clicking on the cell. If the variable is classificatory in nature, coding for each classification is defined under this heading. For example, if data contains male and female both, then the code 1 can be defined for male and 2 for female. These numbers are arbitrary. One can use the coding as 0 and 1 or 2 and 3 as well.

Finally, under the column heading **Scale** the data type of variable needs to be defined by choosing the option; Scale, Nominal, or Ordinal. In SPSS the variable measured on interval or ratio scale is known as Scale variable. Leave all other column entries as default for the time being. After defining all variables the screen will look like as shown in Figure 3.2

After defining all variables in the variable view click on the **Data View** to enter the data of the Table 3.1 column wise for all the variables. The screen shall look like as shown in Figure 3.3.

Table 3.1   Profile Data

Height	Weight	Gender
167	69	M
175	73	M
165	58	F
182	78	M
172	55	F

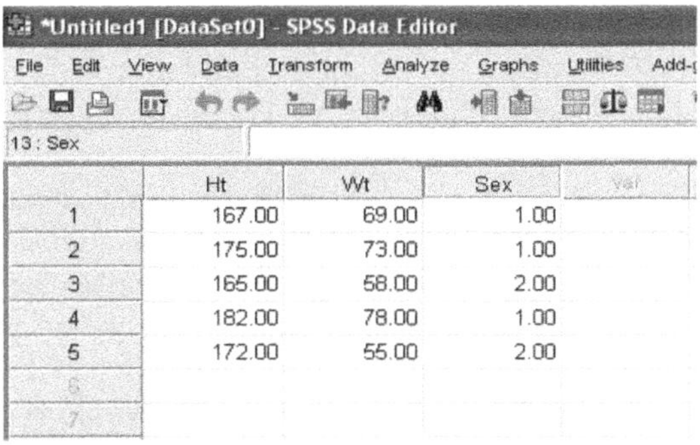

Figure 3.3   Screen showing format of data feeding

Once the data file is ready save it in any location you feel like before using it for any analysis.

ASSUMPTIONS

In using repeated measures design below mentioned assumptions need to be satisfied in order to ensure the validity of findings. In this section the procedure to check these assumptions shall be discussed by using the SPSS software.

a. **Data Type** The independent variable must be categorical and have at least three or more levels. The dependent variable must be measured either on interval or ratio scale.

b. **Independence of Observations** The observations obtained on the subjects should be independent from each other.

c. **Normality** For each level of the independent variable the dependent variable must follow approximately normal distribution and should not have outlier.

d. **Sphericity** The sphericity should not exist among the data. It states that the variances of the differences between all combinations of related groups must be equal. In other words, correlations among the repeated measures are all equal.

The first assumption can be checked by looking to the data and therefore no testing is required for this assumption. The second assumption is known as *randomness* assumption. The randomness of the sample can be tested by using the Run test developed by Abraham Wald and Jacob Wolfowitz. The readers may look into the procedure discussed in the book Verma and Ghufran (2012). In fact the first and second assumptions are design issues and the researcher should plan the study accordingly.

The third assumption is a normality assumption. This assumption shall be tested by using the SPSS software in the next section. The fourth is a Sphericity assumption. This is the most important assumption in repeated measures design. If sphericity assumption is violated some correction is required to be made in the degrees of freedom. Testing of this assumption shall also be discussed by using the SPSS software in the following sections.

Testing Normality

Normality assumption in repeated measures design can be tested by using the SPSS software. Normality of data can be checked from many ways but the most popular way with SPSS is to check it by means of Shapiro–Wilk and Kolmogorov–Smirnov tests. For a normal distribution the values of skewness and kurtosis are zero. Skewness measures symmetricity of data, whereas kurtosis describes the spread of scores around mean value. Any deviation of these statistics from zero indicates nonnormality of data provided they are significant. Using this criteria for testing normality may be problematic if for any variable skewness is significant but kurtosis is not or vice-versa. In that situation it is difficult to draw conclusion about the normality of data. However, Shapiro–Wilk test is a better option in SPSS to test the normality of data because it takes into account both the values of skewness and kurtosis. The SPSS output provides the value of Shapiro–Wilk as well as Kolmogorov–Smirnov tests. The Shapiro–Wilk test is more suitable for the small sample ($N \leq 50$), but it can also be used for sample sizes as large as 2000. On the other hand, Kolmogorov–Smirnov test is more useful for large sample. One of the limitations of these tests is that in case of large sample you are more likely to get significant results. In other words, these tests become significant even for slight deviation from normality in case of large sample.

The procedure of testing normality of data with SPSS shall be shown by means of an example. Let us consider the data on memory recall as shown in Table 3.2 on which the normality assumption shall be tested. The data in this table shows the number of items recalled (out of 10 objects) by the college students in one minute. The subjects have been tested for their memory recall performance in three different time settings (morning, afternoon, and evening).

Let us see how normality can be tested and outliers can be detected by using the SPSS software. To start with, a data file needs to be prepared in SPSS using the data shown in Table 3.2. The procedure of making the data file has been discussed earlier in this chapter. After preparing the data file follow the sequence of commands as shown in Figure 3.4.

After clicking on **Explore** command you will get the screen as shown in Figure 3.5 to select the variables and the options for the analysis. Sift all three variables from the left panel to the "Dependent List" section of the screen. Click on the command **Statistics** and check the 'Outliers' option in the screen as shown in Figure 3.5. This option will detect all the outliers in all the three treatment conditions. Let other options remain selected as default.

After selecting the option for outliers in the screen as shown in Figure 3.5 click on **Continue.** Click on the tag **Plots** in the same screen and then check the 'Normality plots with test' as shown in the Figure 3.6. This option will get you the value of

Table 3.2   Scores on Memory Recall at Different Time

Time		
Morning	Afternoon	Evening
9	4	6
4	5	7
3	6	8
4	4	7
3	5	6
4	3	5
4	1	8
5	6	5
4	4	6
3	9	7
4	5	6
5	6	7

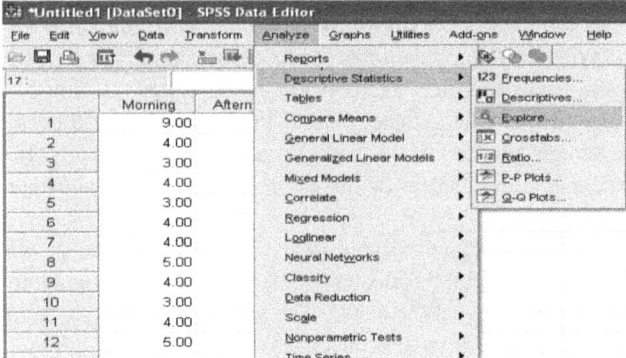

Figure 3.4   Screen for initiating commands for testing normality and identifying outliers

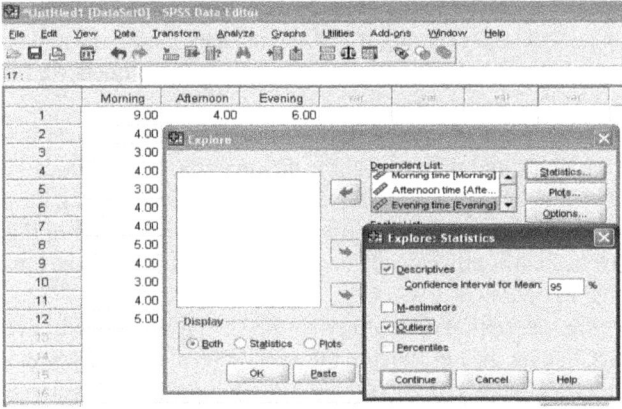

Figure 3.5   Screen showing option for selecting variables and detecting outliers

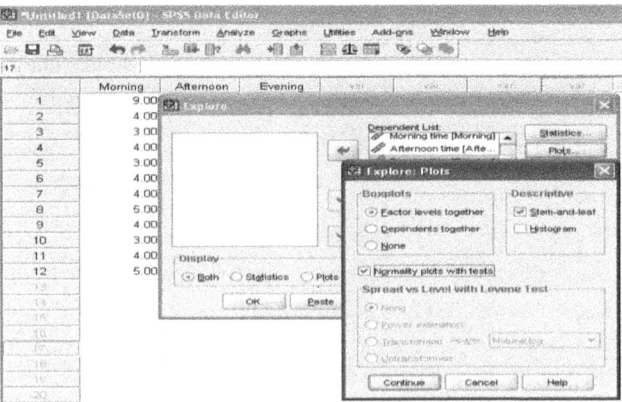

Figure 3.6   Screen showing options for computing Shapiro–Wilk test and the Q–Q plot

Shapiro–Wilk test and Q–Q plot, showing the normality condition of the data graphically. Let other options remain selected as default. Click on **Continue** to go back to the screen shown in Figure 3.5. Click on **OK** to get the output.

There will be many results in the output window of the SPSS but select only three outputs as mentioned below:

1. Tests of normality
2. Q–Q plot
3. Box-plot for identifying outliers

Test of Normality The Table 3.3 shows the Kolmogorov–Smirnov and Shapiro–Wilk test statistics. These tests facilitate you to test normality of the data. Normality exits if these tests are not significant. Thus, if the significance value (p-value) of these tests is more than 0.05, the data is considered to be normal otherwise not. Looking to the values of these tests in Table 3.3, it may be concluded that the data on memory recall obtained in morning testing is nonnormal ($p < 0.05$), whereas normality exists for the data in Afternoon ($p = 0.491$) and Evening ($p = 0.187$) testing. This is so because the Shapiro–Wilk test for the Morning time is significant, whereas for Afternoon and Evening times it is nonsignificant.

Q–Q Plot for Normality Normality of the data can be shown by the graphics also. The normal Q–Q Plot is a graphical way to check the level of normality of data. In Q–Q plot two probability distributions are compared graphically by plotting their quantiles against each other. If distribution of sample data is similar to that of standard normal distribution, the points in the Q–Q plot will lie on the line. The line indicates that the points should fall on it or close by if the data follows normal distribution. If these points deviate much from the line, it indicates nonnormality. The normal Q–Q Plot obtained on the data mentioned in the Table 3.3 is shown in Figure 3.7. In Figure 3.7a the data deviates from the line indicating that the data on memory recall in the Morning group is nonnormal, whereas in the Figure 3.7b and 3.7c the data fall along the line indicating normality of the data in the Afternoon and Evening groups.

Box-plot for Identifying Outliers An outlier is an unusual data which at times affects the findings severely in the study. An outlier may exist in the data set on account of either due to wrong entry or due to extreme observation obtained in the study. For instance in case of data on height, a score of 56 feet is an outlier as it is

Table 3.3   Tests of Normality for the Data On Memory Recall

	Kolmogorov–Smirnov			Shapiro–Wilk		
	Statistics	df	Sig.	Statistic	df	Sig.
Morning time	0.332	12	0.001	0.681	12	0.001
Afternoon time	0.191	12	0.200	0.939	12	0.491
Evening time	0.191	12	0.200	0.906	12	0.187

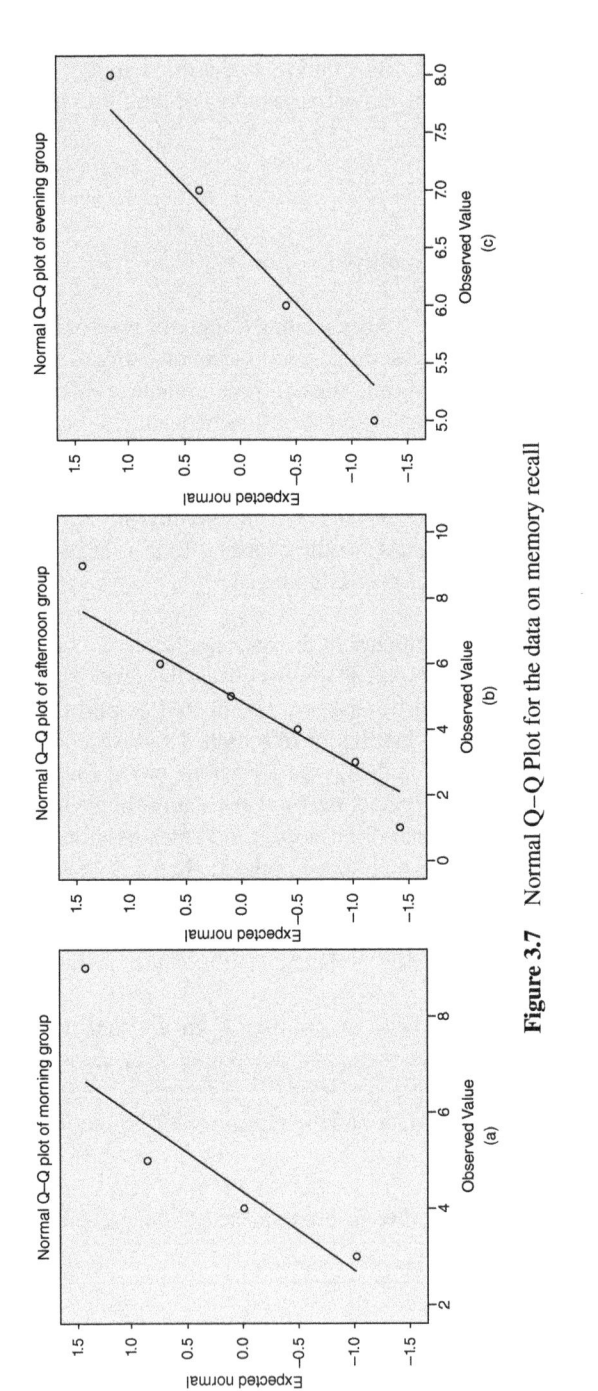

Figure 3.7 Normal Q–Q Plot for the data on memory recall

unusual and is not possible at all and has figured due to recording error whereas if the data is 6.6 ft, it may also be identified as outlier but it has resulted due to extremely tall subject in the study. A researcher must check such outliers before data analysis. If one feels that the data is genuine it may be kept in the study. The outliers in SPSS is identified by using the boxplot. It describes the distribution of data and identifies the outliers if any. Usually any data outside the Mean $\pm$ 2SD or Mean $\pm$ 3SD is taken as outliers. However, SPSS identifies the outlier by using the quartiles. Any data less than $Q_1 - (Q_3 - Q_1)/2$ or more than $Q_3 + (Q_3 - Q_1)/2$ is identified as outlier by the SPSS. You can keep the outlier identified by the SPSS in your study provided you are convinced that such score is genuine and can be obtained by the subjects easily. The Figure 3.8a–c shows the box plot for the scores on memory recall in all the three treatment conditions. It can be seen in Figure 3.8a that the first score in the Morning group is an outlier. You can see that the first data in the Morning group is 9 in the Table 2.1 which is clearly an outlier. The researcher may delete this data from his study.

Testing Sphericity

The sphericity assumption is one of the main assumptions in repeated measures design. Only under the assumption of sphericity the treatment and error variances can be compared by computing the F value. If sphericity assumption is violated then a correction in the degrees of freedom is made while testing the significance of F value. Therefore, it is must that this assumption is tested in the analysis of repeated

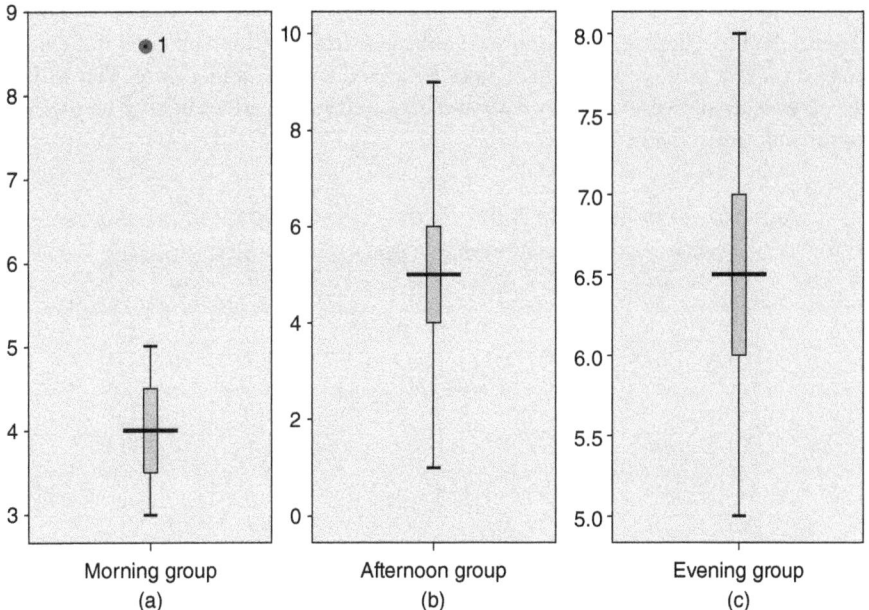

Figure 3.8 Box plot for all three groups of data

measures design. The SPSS software, while solving the repeated measures design, provides the output for testing the sphericity of data. The SPSS uses Mauchly's test in testing the sphericity of data. This test computes a χ^2 statistic which is tested for its significance. Significance of χ^2 indicates that the sphericity exists in the data set. In case the sphericity is significant a statistic epsilon (ε) is computed to indicate its severity. The value of epsilon (ε) ranges from 0 to 1. Lower the value of epsilon (ε) more is the sphericity. The value 1 of epsilon (ε) indicates no sphericity. Depending upon the value of epsilon (ε) a correction is made in the degrees of freedom by using either Greenhouse–Geisser or Huynh–Feldt corrections. If the value of ε is less than 0.75, the Greenhouse–Geisser correction is used for the degrees of freedom, whereas if its value is more than 0.75 the Huynh–Feldt correction is used. Third correction, lower bound reported in the SPSS output is never used by the researchers as it estimates the maximum sphericity in the data set. In case the Mauchly's test is not significant you need not to worry because in that case the sphericity assumption is not violated and the significance of F is tested without making any correction in the degrees of freedom. To show the procedure in SPSS for testing sphericity the data set of Table 3.2 shall be used. Let us see how the sphericity can be tested by using the SPSS software.

After preparing data file follow the below mentioned sequence of commands as shown in Figure 3.9.

After clicking on the **Repeated Measures** command you will get the screen as shown in Figure 3.10 to define the independent and dependent variables. By default the "Within-Subject Factor Name" is written as factor 1. Change this with the name of independent variable. In this case it is *Time*. Type the number of levels as 3 because there are three treatment conditions. Click on **Add**. Now type the name of dependent variable in the "Measure Name" area as *Memory_recall*. Click on **Add**. The name of independent and dependent variables should start from any alphabet and if the name consists of two or more words, they must be joined with the underscore. You will get the screens as shown in Figures 3.10 and 3.11 before and after clicking on the **Add** command, respectively.

Analyze ⟶ General linear model ⟶ Repeated measures

Figure 3.9 Screen for initiating commands for testing sphericity

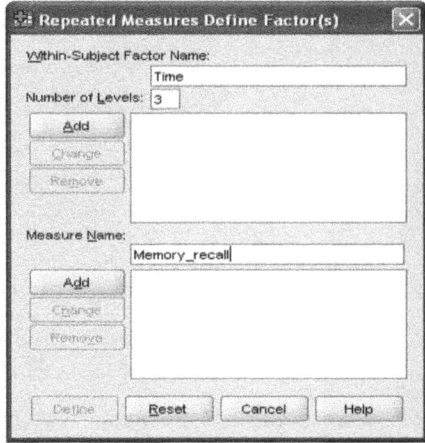

Figure 3.10 Screen showing options for defining variables

Clicking on the command **Define** in the screen shown in Figure 3.11 shall take you to the screen shown in Figure 3.12 for selecting variables and option for the analysis. Select all three variables from the left panel and bring them to the "Within-Subjects Variables" section of the screen. Click on **OK** to get the output.

From the output window of SPSS select only the result of Mauchly's test of Sphericity as shown in the Table 3.4.

Table 3.4 shows that the Mauchly's test is not significant because the p value associated with the χ^2 is 0.495 which is greater than 0.05. Thus, in this case the significance of F value will be tested by assuming sphericity and without doing any correction in the degrees of freedom of treatment and error terms. However, if sphericity assumption is violated the correction is made by using either Greenhouse–

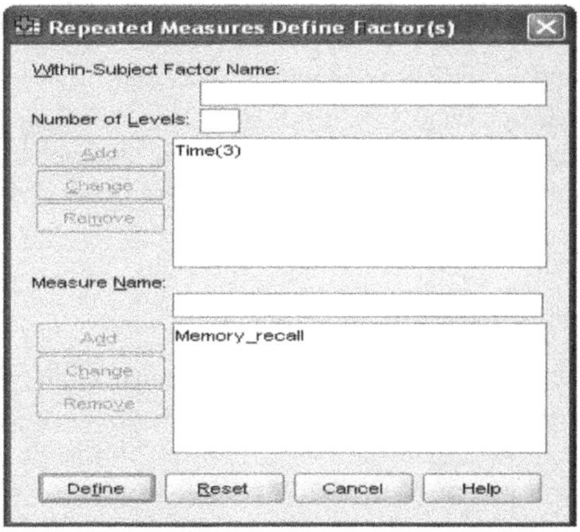

Figure 3.11 Screen showing options for adding independent and dependent variables for analysis

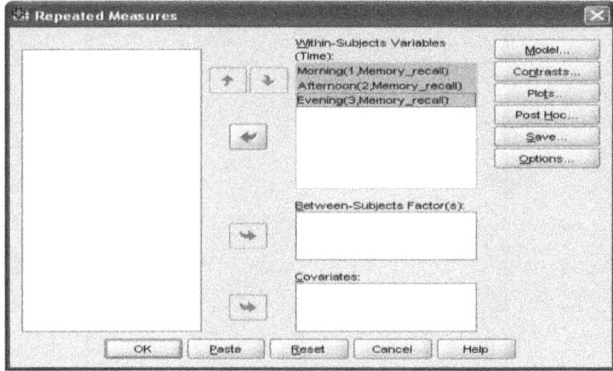

Figure 3.12   Screen showing option for selecting variables for testing sphericity

Table 3.4   **Mauchly's Test of Sphericity**

Measure: Memory_recall							
Within	Mauchly's W	Approx.	df	Sig.	Epsilon		
Subjects Effect		Chi-Square			Greenhouse–Geisser	Huynh–Feldt	Lower-bound
Time	0.869	1.407	2	0.495	0.884	1.000	0.500

Geisser or Huynh–Feldt corrections. The detailed discussion has already been made in Chapter 2 while discussing the repeated measures design.

REMEDIAL MEASURES WHEN ASSUMPTION FAILS

In case if the dependent variable cannot be measured in an interval or ratio scale, repeated measures design cannot be used. However, in that situation one may think of using the Friedman Test which is a nonparametric test used in a situation when the dependent variable is measured on ordinal scale. For detailed discussion the readers may refer Verma and Ghufran (2012).

Transforming Nonnormal Data into Normal

If normality assumption fails, one must think of transforming the variable by using an appropriate transformation so that the transformed variable follows normal distribution. If the transformed variable becomes normal it may be used in the analysis. Usually the transformation like $1/\log(x)$, $1/x$, $1/\sqrt{\log(x)}$, $1/\sqrt[3]{\log(x)}$, $1/x^2$, and $1/(\log(x))^2$ are used to convert nonnormal distribution into normal distribution. However, if the variable can not by transformed to normal even after using different transformations one must think of doing non-parametric analysis.

It can be seen from the Table 3.3 that the distribution of the memory recall scores during morning testing is not normal. Let us see how the distribution of this data set can be transformed into normal distribution using some transformation. All the six suggested transformations have been used to transform the variable x as shown in Table 3.5. These transformations can be applied by using the Excel or SPSS functionality. After applying the transformation six new transformed variables namely B, C, D, E, F, and G have been obtained in the Table 3.5.

By using the procedure discussed above for testing normality in SPSS the normality for all these six transformed variables was tested by using the Shapiro–Wilk test. The results so obtained are shown in the Table 3.6. You can see from this table that the original variable A is not normal because its Shapiro test is significant; however, the variable B (transformed by using the transformation $1/\log(x)$) is normally distributed because the Shapiro test is not significant ($p = .057$) for this variable.

Choice of Design and Sphericity

In repeated measures design if sphericity assumption fails, two of the corrections have been suggested by the Greenhouse–Geisser and Huynh–Feldt to correct the degrees of freedom of treatment and error components. After the correction the F value is tested for its significance. This has been discussed in detail above. However, if the value of epsilon (ε) is very low and the number of subjects are quite large, the multivariate analysis of variance (MANOVA) should be used. In fact the repeated measure design is more efficient than MANOVA design, but under these circumstances the MANOVA is a better option.

Table 3.5 Different Transformation for the Data on Memory Recall in the Morning Group

X	$\dfrac{1}{\log(x)}$	$\dfrac{1}{x}$	$\dfrac{1}{\sqrt{\log(x)}}$	$\dfrac{1}{\sqrt[3]{\log(x)}}$	$\dfrac{1}{x^2}$	$\dfrac{1}{(\log(x))^2}$
A	B	C	D	E	F	G
3.00	2.10	0.33	1.45	1.28	0.11	4.39
3.00	2.10	0.33	1.45	1.28	0.11	4.39
3.00	2.10	0.33	1.45	1.28	0.11	4.39
4.00	1.66	0.25	1.29	1.18	0.06	2.76
4.00	1.66	0.25	1.29	1.18	0.06	2.76
4.00	1.66	0.25	1.29	1.18	0.06	2.76
4.00	1.66	0.25	1.29	1.18	0.06	2.76
4.00	1.66	0.25	1.29	1.18	0.06	2.76
4.00	1.66	0.25	1.29	1.18	0.06	2.76
5.00	1.43	0.20	1.20	1.13	0.04	2.05
5.00	1.43	0.20	1.20	1.13	0.04	2.05
9.00	1.05	0.11	1.02	1.02	0.01	1.10

Table 3.6   Tests of Normality for the Transformed Variable of Shooting Scores in Free Angle Group

	Kolmogorov–Smirnov			Shapiro–Wilk		
	Statistics	df	Sig. (p-value)	Statistic	df	Sig. (p-value)
A	0.332	12	0.001	0.681	12	0.001
B	0.277	12	0.012	0.865	12	**0.057**
C	0.250	12	0.037	0.861	12	0.050
D	0.258	12	0.026	0.858	12	0.046
E	0.268	12	0.017	0.855	12	0.043
F	0.314	12	0.002	0.849	12	0.036
G	0.309	12	0.002	0.854	12	0.041

Bold face indicates that except this variable Shapiro test for all other variables is significant.

SAMPLE SIZE DETERMINATION

In any empirical study sample is used to draw inference about the population characteristics. One of the important issues in such study is to determine the sample size. In general large sample provides more reliable estimate of population characteristics in comparison to small sample, but beyond a certain point it adversely affects the finding.

Size of the sample is usually governed by the cost and accuracy considerations. In other words, it is desired to have maximum statistical power in the experiment with minimum sample. Size of the sample is decided on the basis of any or more of the three factors; firstly, availability of cost; secondly, required target variance; and thirdly, required power in the test. Before we discuss the method of determining sample size, it is important to discuss certain terms used in it.

Important Terms

We shall discuss the following terms first in order to determine the sample size in an empirical study.

Confidence Interval   Population mean can be estimated on the basis of sample by two ways. This can either be done by point or interval estimation. In point estimation a single estimate of the population mean is obtained on the basis of the sample observations, whereas in interval estimation a range of values is obtained in which some confidence is placed that it includes the population mean.

Larger the range more confident we are that it includes the population mean. While testing any hypothesis confidence interval is always computed in the output. Usually 95% or 99% confidence interval is obtained by a researcher. Let us see as to what it means. In estimating mean IQ of students in a college a sample of random students may be obtained. If 95% confidence interval obtained on the basis of sample observations is 80 to 90, let us see what it indicates. It simply means that probability that the confidence interval 80 to 90 includes population mean is 0.95. This fact could be explained like this, if 100 samples are drawn and confidence intervals are constructed for each sample, at least 95 intervals will contain population mean and only 5 or less intervals may not contain it. This fact can be shown by the Figure 3.13 in which if 100 confidence intervals are developed then at least 95 will include the population mean and at the most five intervals may not.

Confidence interval for any parameter is affected by the sample size, population variability and confidence level. The sample size is inversely proportional to the confidence interval. Thus, larger the sample smaller is the confidence interval. If the sample size is equal to the population size, sample mean would be equal to the population mean and the width of the confidence interval would become zero.

Another factor which affects the confidence interval is variability of the population. The confidence interval is directly proportional to the variability of the data in population. Thus, if the population is more homogeneous, the width of the confidence interval would decrease.

Lastly, confidence interval is directly affected by the confidence level you wish to have in estimating a population parameter. Higher the confidence level you require in estimation, larger the confidence interval you will have.

Confidence Level   The confidence level indicates as to how much we are sure that the confidence interval would contain the population parameter we are estimating. It is expressed in terms of percentage and usually taken as 95% or 99% by the researchers. Confidence level is indicated for any confidence interval we develop for estimating the population parameter on the basis of a sample.

Let us see how to interpret the confidence level. Consider that a population consists of IQ scores of students in a college with mean and standard deviation as 80 and 5, respectively, as shown in the Figure 3.14. If IQ scores follow normal distribution, we may conclude that 95% student's IQ would be in the range of 70 to 90 (two sigma

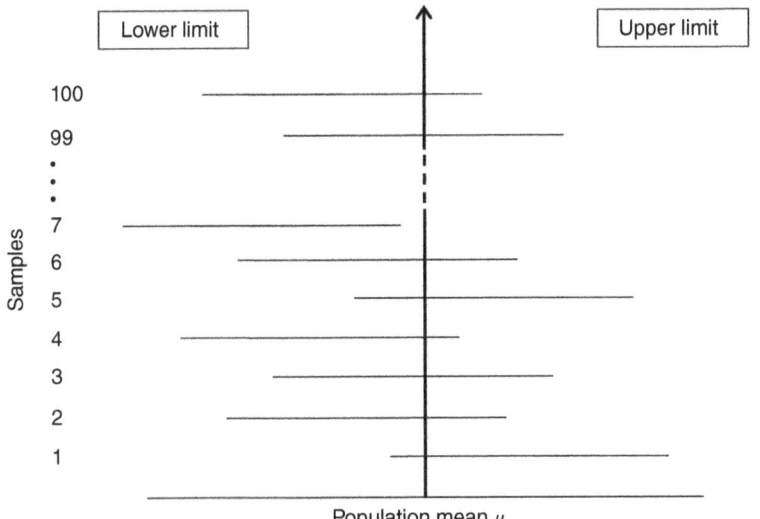

Figure 3.13   Confidence intervals for mean μ

Population: IQ of students

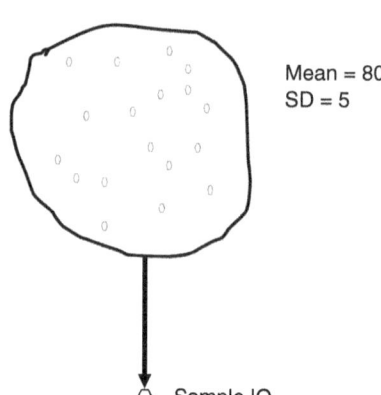

Mean = 80
SD = 5

Sample IQ

Figure 3.14   Showing IQ scores in a population

limits). This can also be interpreted as, if a sample student is randomly selected from the population and if his IQ is in the range of 70–90, we shall be 95% confident that he belongs to the population.

Power of the Test   Power of a statistical test refers to its ability in accepting the claim if it is correct in a hypothesis testing experiment. In other words, if a research hypothesis is correct, the null hypothesis should be rejected while testing.

To understand power it is essential to understand the errors committed on the part of a researcher in testing a hypothesis. In any hypothesis testing, claim of a new proposition is tested by negating the null hypothesis. For instance, consider a situation in which a drug manufacturer launches a new drug claiming to cure tuberculosis patients in a reduced time period in comparison to the existing drug. An experiment may be conducted to test the claim. Here the whole focus of the researcher is to test whether the null hypothesis can be rejected at a desired level of significance. While testing a null hypothesis a researcher may commit two kinds of errors known as Type I and Type II. Type I error refers to rejecting a null hypothesis when it is true. On the other hand, Type II error is said to occur if null hypothesis is not rejected when alternative hypothesis is true. Probability of Type I and Type II errors are known as α and β, respectively. Type I error is known as consumer's risk, whereas Type II is known as producer's risk. Type I error is considered to be more serious than Type II error. It is because of the fact that wrongly rejecting a null hypothesis leads to wrongly accepting an alternative hypothesis (claim).

The power of a test is computed by $1 - \beta$. This indicates the probability of rejecting a null hypothesis if the alternative hypothesis is correct. A researcher needs to decide in advance as to how much power he needs to have in his experiment as this affects the size of the sample in empirical studies.

Sample Size Determination on the Basis of Cost

In a study sample size can be decided on the basis of the cost. In that case one must have the knowledge about the population, cost of the equipments, and other resources used in the study. In selecting the size of a sample in any empirical study, the trade off is between the cost of the project and power of the test. In most of the research studies limited finance is available; hence, researcher may use cost consideration for deciding the sample size and see how much power can be achieved with the sample size selected in the study. The below mentioned model for sample size determination can be used by a researcher if cost is the main consideration.

If C is the total cost available for the study, the required sample size can be determined by the following formula:

$$C = a + nc_1 + \frac{C}{10} \tag{3.1}$$

Where, "a" is the overhead cost, "c_1" is the cost of evaluating an experimental unit, and "n" is the sample size. The overhead cost refers to one time expenditure like the cost of equipments, stationary, printing, computing, and so on. The value of c_1 is the cost of per unit evaluation. This includes the cost of monitoring the data collection, cost of field investigator, and cost of motivating respondents. Here the cost component $C/10$ refers to the contingency cost.

Consider that in a study, cost of per case evaluation for the field investigator is \$ 30, the cost of motivating per respondent is \$ 5, and the cost of the project coordinator per case is \$ 5, then the cost of evaluating per unit (c_1) would become \$ 40. Further, if the over head cost (a) is \$ 5000 and the total cost of the project (C) is \$ 10,000, then by using the equation (3.1) the sample size (n) would be 100, that is,

$$C = a + nc_1 + \frac{C}{10}$$

$$\Rightarrow 10,000 = 5000 + n \times 40 + \frac{10,000}{10}$$

$$\Rightarrow n = 100$$

Sample Size Determination on the Basis of Accuracy Factor

By accuracy in an empirical study we mean as to how much power we wish to have in hypothesis testing study, at what level of significance we desire to test a hypothesis, and how much variation we wish to estimate with the true value of the population characteristics. We will show the concept involved in sample size determination in estimating a population mean.

Sample Size in Estimating Mean   As per the central limit theorem if a random sample is drawn from a population having mean and standard deviation as μ and σ, respectively, the distribution of the sample mean ($\bar{x}$) follows a normal distribution with mean μ and standard deviation $\sigma/\sqrt{n}$ provided the sample is sufficiently large ($n \geq 30$). Thus, by using the normal distribution property the 95% confidence interval as given below would estimate the population mean within it.

$$\left(\bar{x} - 2\frac{\sigma}{\sqrt{n}}, \bar{x} + 2\frac{\sigma}{\sqrt{n}} \right)$$

If we wish to estimate population mean within "d" units of the true mean at 95% confidence level we would solve

$$\left(\bar{x} + 2\frac{\sigma}{\sqrt{n}} \right) - \left(\bar{x} - 2\frac{\sigma}{\sqrt{n}} \right) = 2d$$

$$\Rightarrow 4\frac{\sigma}{\sqrt{n}} = 2d$$

$$\Rightarrow n = \frac{4\sigma^2}{d^2}$$

If a researcher wishes to have a 95% confidence interval for the mean IQ of the students and desires that the estimate should be within 2 scores of the true value, where standard deviation of the IQ is 8, then the sample size required for the study would be 64 as shown below.

$$n = \frac{4\sigma^2}{d^2} = \frac{4 \times 8^2}{2^2} = 64$$

Sample Size in Hypothesis Testing In hypothesis testing experiment determination of sample size depends upon the power of the test a researcher wishes to have, effect size, and the significance level at which a hypothesis is desired to be tested. For determining the sample size in repeated measures design, readers may use the program GLIMMPSE (URL: http://glimmpse.samplesizeshop.org/). This is a free internet-based program.

For determining the power of test and determining the sample size, readers may also use the resources available on the below mentioned websites as well:

http://ps-power-and-sample-size-calculation.software.informer.com
http://www.ncss.com/software/pass/
http://www.epibiostat.ucsf.edu/biostat/sampsize.html#ttest
http://www.studysize.com
http://www.poweranalysis.com

EXERCISE

3.1. What action a researcher should take if the normality and sphericity assumptions do not hold in the repeated measures design?

3.2. Discuss the procedure of making data file in SPSS.

3.3. What is outliers and how it affects findings in a study? What strategy is adopted in detecting outlier and why?

3.4. What are the various methods of testing normality of data? What will you do if the data is nonnormal in solving analysis of variance?

3.5. List down the assumptions that are required to be satisfied in repeated measures design.

3.6. What is sphericity and how it is computed? Discuss various corrections that are applied if sphericity assumption is violated in repeated measures design.

ASSIGNMENT

3.1. Following are the marks of students in different subjects. Test normality of these data and identify the outliers if any. Check the significance of skewness and kurtosis of these data and interpret your findings. Apply different transformations to check whether nonnormal data can be converted into normal data.

SN	Marks of the Students		
	Maths	English	Science
1	55	45	19
2	58	52	47
3	56	49	59
4	48	52	22
5	92	72	67
6	65	45	56
7	62	49	52
8	66	48	51
9	53	47	50
10	48	45	61

3.2. Following are the scores on accuracy test obtained on the subjects under three different environmental conditions. Do the following:

a. Check normality of three sets of repeated measures.

b. Try converting nonnormal data into normal by using the transformation.

c. Test the significance of skewness and kurtosis of these data at 5% level and interpret your findings.

d. Identify the outliers if any in the data set by obtaining the boxplot.

e. Test sphericity of the data.

	Data on Accuracy Test Under Different Environment		
SN	Environment		
	Hot	Cold	Humid
1	26	39	34
2	32	36	22
3	32	35	20
4	20	38	21
5	29	20	19
6	31	34	34
7	32	32	32
8	25	30	35

BIBLIOGRAPHY

Algina. Generalized eta and omega squared statistics: Measures of effect size for some common research designs. Psychol Methods 2003;8:434–447. DOI: 10.1037/1082-989x.8.4.434.

Cochran WG. *Sampling Techniques*. 2nd ed. New York: John Wiley and Sons, Inc.; 1963.

D'Agostino RB. Tests for the Normal Distribution. In: D'Agostino RB, Stephens MA, editors. *Goodness-of-Fit Techniques*. Marcel Dekker: New YorkISBN: 0-8247-7487-6; 1986.

Filliben JJ. The probability plot correlation coefficient test for normality. Technometrics 1975;17(1):111–117. DOI: 10.2307/1268008JSTOR 1268008.

Green SB, Salkind NJ. *Using SPSS for Windows and Macintosh: analyzing and understanding data*. 6th ed. Boston; 2010 Prentice HallISBN: 978-0-205-02040-9; 0000.

Hinton PR, Brownlow C, McMurray I. *SPSS Explained*. Routledge; 2004.

Kish L. *Survey Sampling*. New York: John Wiley and Sons, Inc.; 1965.

Mardia KV. Measures of multivariate skewness and kurtosis with applications. Biometrika 1970;57:519–530.

Marsaglia G, Tsang WW, Wang J. Evaluating Kolmogorov's distribution. J Stat Software 2003;8(18):1–4.

Mauchly JW. Significance test for sphericity of a normal n-variate distribution. Ann Math Stat 1940;11(2):204–209. DOI: 10.1214/aoms/1177731915JSTOR 2235878.

Miaoulis G, Michener RD. *An Introduction to Sampling*. Dubuque, Iowa: Kendall/Hunt Publishing Company; 1976.

Park, H. M. (2002–2008). Univariate Analysis and Normality Test Using SAS, Stata, and SPSS. [working paper]. Retrieved 26 February 2014.

Pearson ES, Hartley HO, editors. *Biometrika Tables for Statisticians 2*. Cambridge University Press; 1972. p 117–123 Tables 54, 55 ISBN: 0-521-06937-8.

Razali N, Wah YB. Power comparisons of Shapiro-Wilk, Kolmogorov-Smirnov, Lilliefors and Anderson-Darling tests. J Stat Modell Anal 2011;2(1):21–33 Retrieved 5 June 2012.

Royston P. Approximating the Shapiro–Wilk W-test for non-normality. Stat Comput 1992;2(3):117–119. DOI: 10.1007/BF01891203.

Shorack GR, Wellner JA. *Empirical Processes with Applications to Statistics*. Wiley; 1986. p 239 ISBN: 047186725X.

Shuttleworth, M. (2009). Repeated Measures Design. Experiment-resources.com. Retrieved 2013-09-02.

Sphericity. Laerd Statistics. https://statistics.laerd.com/spss-tutorials/two-way-repeated-measures-anova-using-spss-statistics.php

Spiegelhalter DJ. An omnibus test for normality for small samples. Biometrika 1980;67:493–496. DOI: 10.1093/biomet/67.2.493.

Sudman S. *Applied Sampling*. New York: Academic Press; 1976.

Székely GJ, Rizzo ML. A new test for multivariate normality. J Multivariate Anal 2005;93:58–80.

Thadewald T, Büning H. Jarque–Bera test and its competitors for testing normality – a power comparison. J Appl Stat 2007;34(1):87–105. DOI: 10.1080/02664760600994539 Retrieved 5 June 2012.

Thode HC Jr. *Testing for Normality*. New York: Marcel Dekker, Inc.; 2002. p 479 ISBN: 0-8247-9613-6.

Verma JP Ghufran M. (2012) Statistics for Psychology: A Comprehensive Text. McGraw Hill Education Private Limited

How do I interpret the Shapiro–Wilk test for normality? JMPRetrieved March 24, 2012. 2004.

4

ONE-WAY REPEATED MEASURES DESIGN

INTRODUCTION TO DESIGN

In one-way repeated measures design effect of one independent factor on some criterion variable is investigated by repeatedly testing all subjects in each treatment. These levels of factor are referred as treatments. Consider an experiment in which it is required to investigate the effect of different caffeine concentration on blood lactate. One may take three levels of caffeine concentration in the experiment. Here, caffeine is an independent factor and its different concentrations are treatments. Since in this design subjects are tested under each treatment to avoid individual variation, each subject needs to be tested for their blood lactate level after consuming caffeine with different concentration. Sometimes, the researcher may also like to see the impact of training on the performance over a period of time. For instance, one might be interested to know the effect of endurance training program on lungs capacity in different durations. The researcher may test the subjects before starting the experiment and after 2 weeks, 4 weeks, and 6 weeks of endurance training. The purpose of such experiment is to investigate the behavioral trend of subjects in relation to the criterion variable over a period of time. Sometimes, this design is also known as time series design and is similar to the repeated measures design. Since each subject is repeatedly tested under different treatment conditions, this design is also known as *within-group design* or *within-subjects design*.

Repeated Measures Design for Empirical Researchers, First Edition. J. P. Verma.
© 2016 John Wiley & Sons, Inc. Published 2016 by John Wiley & Sons, Inc.

ADVANTAGE OF ONE-WAY REPEATED MEASURES DESIGN

This design has the following benefits:

1. Less number of subjects is required for the study due to which an experimenter can have more control in the experiment.
2. Since same subjects are tested in all treatments, subjects serve their own control due to which experimental error is reduced.
3. The repeated measures design is more efficient in comparison to that of the independent measures design because a part of the treatment group's variability is explained by the subjects thereby experimental error is reduced.
4. This design is very sensitive in nature due to which a slight change in the criterion variable due to the manipulation of treatment conditions can be easily detected.
5. This design allows the experimenter to study the behavior of subjects due to intervention of treatment over the period of time.

WEAKNESS OF REPEATED MEASURES DESIGN

Due to carryover effect performance of the subjects may be affected in different treatment conditions. In other words, participation in one treatment may affect the performance of subjects in other treatment groups resulting creation of a confounding extraneous variable that varies with the independent variable. The carryover effect may be because of the fatigue experienced or due to learning in earlier treatment conditions. Another weakness of this design is that if a subject is tested in all treatments in a specific sequence an order effect may be generated which affects the performance of the subjects. Further, if the number of treatments are large, subjects may loose interest in the experiment as they might feel bore which in turn affects the outcomes in the study.

APPLICATION

There may be numerous application of this design in different disciplines. Initially, the design of experiment was developed for agricultural research in which basic experimental units were plots having fixed variability. In such experiment, behavior of each plot does not differ over experimental period. But this is not the case with studies where experimental units are human beings or animals because performance of subjects not only is affected by manipulating treatments but is also affected by the mood, temperament, and motivation of the subjects. Thus, it is important to control the effect of different subjects used in different treatment groups in an independent measures design. In repeated measures design, this individual variation can be controlled by testing each subject in all treatments. Thus, in all those studies where experimental units are human beings or animals, this design is more efficient. Following are some of the specific situations where this design can be used to test desired research questions.

1. In a clinical experiment for investigating body's reaction to drug, the drug efficacy can be tested by taking hourly blood samples for 12 hours after its administration. Thus, a series of data is obtained on an individual for measuring blood concentration levels.

2. A coach may like to compare recovery pattern of soccer players under light exercise, autogenic relaxation, and underwater exercise to find as to which intervention is the best. The recovery pattern of same group of subjects is studied under each intervention after three soccer match of similar level.

3. A physiologist may study an intervention of pranayama in the relief of asthma by measuring the subject's forced expiratory volume (FEV) at weekly intervals to assess the efficacy of intervention.

4. In order to study the taste of three different pizza (pan, cheese, and chicken) on the youngsters a market research analysts may offer a group of subjects to eat these pizza free of costs one by one. After consuming each pizza, subjects may be asked to rate their tastes on a 10-point scale.

LAYOUT DESIGN

The layout of the design depends upon whether the levels of the independent variable are the different treatments or the time period. Accordingly two different kinds of layout are as follows:

Case I: When the Levels of Within-Subjects Variable are Different Treatments

In this layout design, each subject is tested under different treatments. In such studies where this design is used, the researcher must ensure that there is no learning effect due to the treatments. To done away the learning effect, sufficient time gap should be given in between implementing any two treatments. Another issue in such studies is that of order effect. Due to implementing treatment conditions in a specific sequence the subject's performance may be affected. The performance of the subjects in the second treatment condition may be affected due to the learning effect or the exhaustion effect experienced by the subjects during the first treatment condition. Since the same subjects are being tested in different treatment conditions, this may affect the subject's performance adversely if they do not find it interesting or gets fatigue during treatment implementation.

To done away the order effect, counterbalancing is used in the design. It is done by dividing the sample into number of groups; say three, if there are three treatments. The first group undergoes the first treatment thereafter second and then the third one, whereas the second group may undergo into the second treatment first and then the first and thereafter the third one. Similarly, the order of the third group may be the third treatment first then the second and thereafter the first one.

In the pizza example discussed above if six subjects are taken in the sample, then the layout of the one-way repeated measures design can be shown by the Figure 4.1. In this layout, on the first day of testing the subjects S1 and S2 consume pan pizza, the

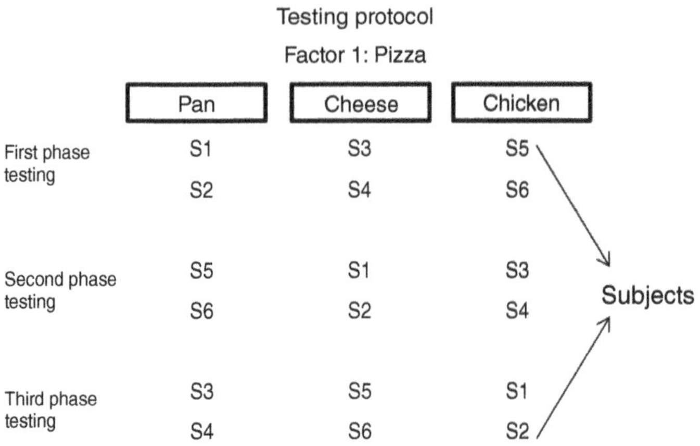

Testing protocol

Factor 1: Pizza

Pan	Cheese	Chicken

First phase testing

| S1 | S3 | S5 |
| S2 | S4 | S6 |

Second phase testing

| S5 | S1 | S3 |
| S6 | S2 | S4 |

Subjects

Third phase testing

| S3 | S5 | S1 |
| S4 | S6 | S2 |

Figure 4.1 Layout of the one-way repeated measures design having levels as different treatments

subjects S3 and S4 eat cheese pizza, and the subjects S5 and S6 get chicken pizza. Similarly, the testing protocol for the subjects during second and third testing has been shown.

Case II: When the Levels of Within-Subjects Variable are Different Time Durations

In this layout design, all subjects are repeatedly tested at different time points during treatment. Here, performance of subjects at a particular time point is the result of cumulative learning during experiment. In such studies, researcher is interested to know the behavior of subjects due to the treatment over the period of time. If in the drug efficacy example discussed earlier if five subjects are tested for blood concentration before administration of drug and after 2 hours, 4 hours, and 6 hours during treatment then the layout of this repeated measures design can be as shown by the Figure 4.2.

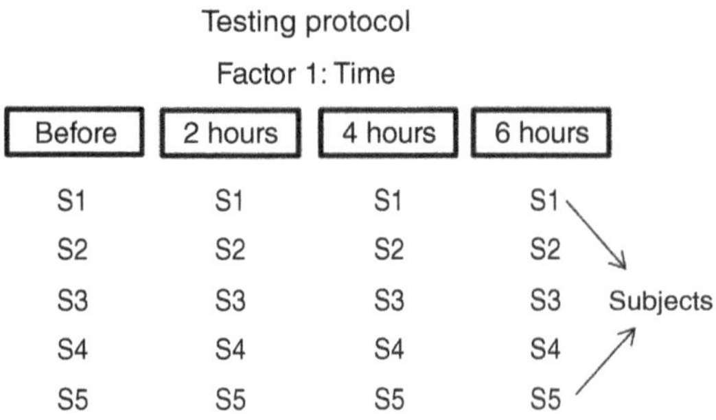

Testing protocol

Factor 1: Time

Before	2 hours	4 hours	6 hours
S1	S1	S1	S1
S2	S2	S2	S2
S3	S3	S3	S3
S4	S4	S4	S4
S5	S5	S5	S5

Subjects

Figure 4.2 Layout of the one-way repeated measures design having levels as time durations

ILLUSTRATION 77

STEPS IN SOLVING ONE-WAY REPEATED MEASURES DESIGN

One should follow the below-mentioned steps for solving one-way repeated measures design with the SPSS software:

1. Test the assumption of normality as discussed in Chapter 3.
2. Describe the layout design.
3. Write research questions which need to be investigated.
4. Write null hypothesis to be tested.
5. Decide familywise error rates (α) at which the hypothesis is required to be tested.
6. Use SPSS procedure to generate the following outputs:
   a. Descriptive statistics including mean and standard deviation
   b. Mauchly's test of sphericity including estimated value of epsilon
   c. Table of F value for testing within-subjects effect
   d. Pairwise comparison of means
   e. Means plot
7. Test sphericity assumption.
8. If sphericity is significant apply correction (Greenhouse–Geisser or Huynh–Feldt) in the degrees of freedom of SS between treatments and within treatments and test the significance of F value at the corrected degrees of freedom otherwise test the significance of F by assuming sphericity.
9. If F ratio is significant apply paired t tests for pairwise comparisons among group means by applying the Bonferroni correction.
10. Report the findings.

The detail procedure in solving repeated measures design shall be explained by means of the below-mentioned illustration.

ILLUSTRATION

A researcher used an intervention of meditation program on 10 randomly selected male subjects to see its impact on the performance of reasoning ability in different durations. The subjects were asked to solve similar level of questions to test their reasoning ability. Performance of the subjects was obtained on zeroth day, 2nd week, 4th week, and 6th week. The scores so obtained are shown in Table 4.1. We shall discuss the procedure involved in solving the one-way repeated measures design with the help of this illustration by using the level of significance as 0.05

Testing Assumptions

In this repeated measures design following assumptions need to be satisfied:

Table 4.1   Data on Reasoning Ability Measured at Different Time
Points During Intervention Program With Meditation

Subjects	Zeroth day	2nd Week	4th Week	6th Week
S1	31	36	35	37
S2	31	34	34	35
S3	32	31	37	35
S4	30	32	36	35
S5	34	33	37	37
S6	35	34	36	38
S7	36	31	31	38
S8	36	35	30	40
S9	32	31	35	36
S10	33	32	34	36

a. **Data Type** The independent variable must be categorical and should have at least three levels and the dependent variable must be measured either on interval or ratio scale.

b. **Independence of Observations** The subjects must be randomly selected and tested independently to each other.

c. **Normality** For each level of the independent variable, the dependent variable must follow approximately normal distribution and should not have outlier.

d. **Sphericity** The sphericity should not exist among the data. This means that the variances of the differences between all combinations of related groups must be equal. In other words, correlations among the repeated measures are all equal.

Let us see whether all these assumptions have been met in this illustration. Here, the independent variable is time (with four levels) and the dependent variable is reasoning ability which is measured on interval scale; hence, the first assumption is satisfied. Since the sample has been randomly selected and the observations have been independently obtained on the subjects, the second assumption is also fulfilled.

The data given in Table 4.1 follows normal distribution because the Shapiro–Wilk test is not significant for any of the four groups of the data as shown in Table 4.2. Further, there is no outlier in the data (i.e., no data is outside the range of $\bar{x} \pm 2s$). The readers are advised to test this assumption by using the procedure shown in Chapter 3. Thus, the assumption of normality is also satisfied.

Finally, the assumption of sphericity will be tested by using the SPSS while generating the output in solving this design. It can be seen from the Table 4.4 that the sphericity has been violated; hence, the correction shall be applied for testing the significance of F value. This aspect shall be discussed later while discussing the findings.

Layout Design

In this study, effect of meditation duration is to be investigated on the reasoning ability and the design is a repeated measure where all the subjects are tested at different time points; hence, the layout of the design in the study would be as shown in Figure 4.3.

ILLUSTRATION 79

Table 4.2   Tests of Normality for the Data on Reasoning Ability

	Kolmogorov–Smirnov			Shapiro–Wilk		
	Statistics	df	Sig.	Statistics	df	Sig.
Zeroth day	0.178	10	0.200	0.924	10	0.393
2nd week	0.192	10	0.200*	0.905	10	0.246
4th week	0.216	10	0.200*	0.879	10	0.128
6th week	0.166	10	0.200*	0.902	10	0.228

* Significant at 5% level

Testing protocol

Factor 1: Time

Zero day	2nd week	4th week	6th week
S1	S1	S1	S1
S2	S2	S2	S2
.	.	.	.
.	.	.	.
.	.	.	.
S9	S9	S9	S9
S10	S10	S10	S10

Subjects

Figure 4.3   Layout of the design for the study shown in the illustration

In the repeated measures design, subjects serve their own control and part of the experimental variability is explained by the subjects resulting reduction in error variance. This enhances the efficiency of repeated measures design in comparison to that of the independent measures design.

Distribution of Variation and Degrees of Freedom

Let us see how the total variation in this design is distributed among different components. If you look into the data of Table 4.1, the total variability (SS) is because of variation due to the data in four groups (SS_{Time}) and variation due to within each of the four groups (SS_{Within}). Since part of the within group variability has been explained by the subjects, this SS_{Within} has been further broken into $SS_{Subjects}$ and SS_{Error}. Thus, in repeated measure design the error (SS_{Within}) has been further reduced to SS_{Error}. This reduction of error variance improves the efficiency of repeated measures design. Reduction of error variance causes increase in the value of F and therefore repeated measures design is more sensitive in detecting even the little difference observed in the dependent variable due to the change in the levels of the independent variable.

In this illustrated repeated measure design, the total degrees of freedom, that is, $N-1(=39)$ has been partitioned into $r-1(=3)$ df due to variation between groups (Time) and $N-r(=36)$ df due to variation within groups. Since SS within groups is

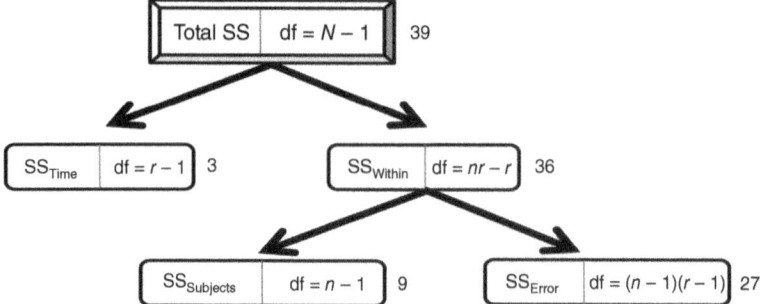

Figure 4.4   Scheme of distributing total sum of squares and degrees of freedom

further divided into SS due to subjects and SS due to error, the degrees of freedom for SS within groups ($nr - r = 36$) is further split into $n - 1(=9)$ df due to subjects and $(n - 1)(r - 1)[=27]$df due to error.

The distribution of Total SS into different components along with their degrees of freedom has been shown in Figure 4.4.

Hypothesis Construction

While planning a study, a researcher must mention the research question to be investigated. In this illustration, the research question is to test whether the duration of meditation training affects the reasoning ability. To investigate this question, a null hypothesis needs to be stated. The null hypothesis is a via media of testing the research hypothesis. Since research hypothesis is always tested by contradicting the null hypothesis, the researcher is always interested to test whether the null hypothesis can be rejected or not on the basis of the given sampled data. In statistics, the research hypothesis is also known as alternative hypothesis; hence, in this study the following null hypothesis shall be tested:

$$H_0 : \mu_{Zero\_day} = \mu_{2nd Week} = \mu_{4th Week} = \mu_{6th Week}$$

against the alternative hypothesis that at least one of the group mean differs.

Level of Significance

Level of significance in testing the significance of F ratio for comparing within-subjects levels is normally taken as either 0.05 or 0.01. In repeated measures design, there is no provision of post hoc test and therefore if the F ratio for within-subjects effect is significant then the paired t-test is used for pairwise comparison of means by using the Bonferroni correction. This correction suggests that the significance of paired t-test should be tested at α/k level of significance, where α is the family wise error rate at which the hypothesis is tested and k is the number paired comparisons required to be made in the study. By using the Bonferroni correction in SPSS, this correction is automatically taken care of by increasing p value.

ILLUSTRATION 81

Figure 4.5 Screen for initiating commands for single factor repeated measures design

Solving One-Way Repeated Measures Design Using SPSS

SPSS software can be used for solving this design. To analyze this design, a data file needs to be prepared in SPSS. Procedure for preparing the data file has been explained in Chapter 3. Those who are using the SPSS for the first time are advised to go through this chapter first before following the procedure discussed here. Once the data file is created, follow the sequence of commands mentioned below as shown in the screen in Figure 4.5.

Analyze ⟶ General linear model ⟶ Repeated measures

After clicking on **Repeated Measures** command you will get the screen as shown in Figure 4.6 to define the independent and dependent variables. By default the "Within-Subject Factor Name" is written as factor 1. Replace this by *Time* because Time is the independent variable in this illustration. Type the number of levels as 4 because there are four time periods in which the data has been obtained. Click on **Add**. Type the name of dependent variable in the "Measure Name" area as *Reasoning_ability*. Click on **Add**. Name of the independent and dependent variables

Figure 4.6 Screen showing options for defining dependent and independent variables

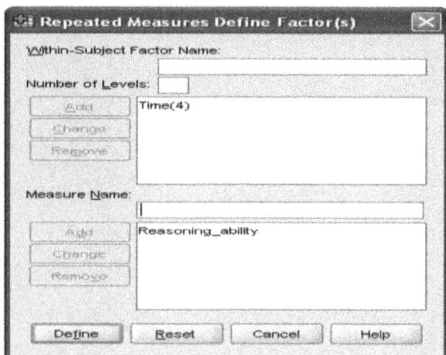

Figure 4.7   Screen showing options for adding independent and dependent variables for analysis

should always start from alphabet and if it consists of two or more words then they should be joined with the underscore. You will get the screens as shown in Figures 4.6 and 4.7 before and after clicking on **Add** command respectively.

Clicking on the command **Define** in the screen shown in Figure 4.7 shall take you to the screen shown in Figure 4.8 for selecting the within-subjects variables. Select all the four variables from the left panel and bring them to the "Within-Subjects Variables" section of the screen. Click on the command **Plots** and transfer the variable *Time* from the "Factor" section into the "Horizontal Axis" area. Click on **Add** to get the means plot in the output.

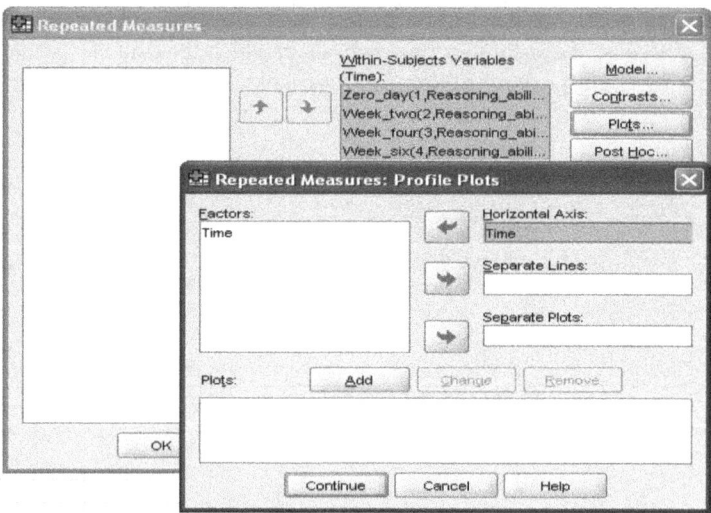

Figure 4.8   Screen showing option for selecting within-subjects variables and obtaining means plot

ILLUSTRATION 83

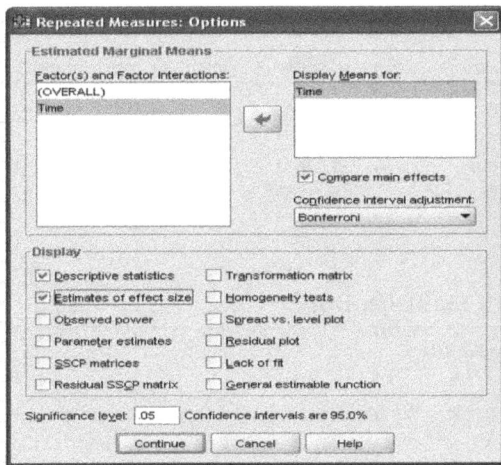

Figure 4.9 Screen showing option for computing descriptive statistics and pairwise comparison of means using the Bonferroni correction

Click on **Continue** for selecting further option in the design as shown in Figure 4.9. In this screen transfer, the variable *Time* from the "Factor(s) and Factor Interactions" section to the "Display Means for" section. Check the option 'Compare Mean effects' and 'Estimates of the effect size'. Select the 'Bonferroni' correction by clicking on the sign ▼ mentioned on the "LSD(none)" tag. Check the option 'Descriptive statistics' for computing mean and standard deviation in each group. Ensure that the level of significance is selected as 0.05. In fact, it is selected by default. Let all other options remain selected as it is. Click on **Continue** to get back the screen as shown in Figure 4.8. Click on **OK** to get the outputs.

SPSS Output and Interpretation

Lots of results are generated in the output window of SPSS, but only the below mentioned outputs shall be selected for interpretation. Right click the mouse over the output which is required to be selected, copy it and paste it in your word document.

- Descriptive statistics
- Mauchly's test of sphericity
- *F* table for testing within-subjects effects
- Table for pairwise comparison of means
- Marginal means plot

Descriptive Statistics The first output refers to the descriptive statistics as shown in Table 4.3. These results can be shown in the project in order to have an idea about the central location and measure of spread in the data of different groups. These values of mean may be used to compare the marginal means graphically.

Table 4.3   Descriptive Statistics

	Mean	SD	N
Zero_day	33.0000	2.16025	10
Week_two	32.9000	1.79196	10
Week_four	34.5000	2.36878	10
Week_six	36.7000	1.63639	10

Table 4.4   Mauchly's Test of Sphericity[a]

Measure: Reasoning_ability

Within Subjects Effect	Mauchly's W	Approx. Chi-Square	df	Sig.	Epsilon[a] Greenhouse–Geisser	Huynh–Feldt	Lower-bound
Time	0.062	21.441	5	**0.001**	0.546	0.650	0.333

[a]Design: Intercept. Within-subjects Design: Time
Bold number indicates that the effect is significant at 5% level.

Testing Sphericity   Table 4.4 is the output for testing the sphericity and shows the estimates of Epsilon required for correcting the degrees of freedom in testing the significance of F value. Mauchly's statistic follows approximately chi-square distribution; hence, the significance of the chi-square is tested for testing the sphericity. It can be seen from this table that Mauchly's test is significant because the p value associated with the chi-square statistic is 0.001 which is less than 0.05. Since Mauchly's test is significant, sphericity assumption violates in this illustration. This requires some corrections to be made in the degrees of freedom of the treatment and the error components before testing the significance of F value.

Correction for Sphericity   The estimated value of epsilon (ε) indicates the severity of sphericity violation. The value of ε can be in between 0 to 1. Lesser the value of ε greater is the violation in sphericity. If the value of epsilon (ε) is less than 0.75, use Greenhouse–Geisser correction otherwise use Huynh–Feldt correction. In this illustration, the value of ε estimated by the Greenhouse–Geisser is 0.546 as shown in Table 4.4, which is less than 0.75; hence, the Greenhouse–Geisser estimate shall be used for correcting the values of degrees of freedom. If there were no sphericity the degrees of freedom for the treatment (Time) and Error would have been 3 and 27 as shown in Table 4.5. Since the sphericity exists, these degrees of freedom shall not be used to find the table value of F or its associated p value. Instead, the table value of F will have to be seen at the degrees of freedom (1.637, 14.730) instead of (3, 27). In SPSS, the significance of F value is tested by means of p value and therefore you can notice that the p value differs in a situation where the Greenhouse–Geisser correction is used ($p = 0.009$) and where sphericity is assumed ($p = 0.001$). Thus, the Greenhouse–Geisser correction simply changes the value of p and nothing more.

ILLUSTRATION

85

Table 4.5 *F* **Table for Testing Significance of Within-subjects Effects**

Measure: Reasoning_ability

Source		Type III SS	df	Mean square	F	Sig.	Partial eta squared
Time	Sphericity assumed	94.475	3	31.492	7.281	0.001	0.447
	Greenhouse–Geisser	94.475	**1.637**	57.725	**7.281**	**0.009**	0.447
	Huynh–Feldt	94.475	1.951	48.423	7.281	0.005	0.447
	Lower-bound	94.475	1.000	94.475	7.281	0.024	0.447
Error (Time)	Sphericity assumed	116.775	27	4.325			
	Greenhouse–Geisser	116.775	**14.730**	7.928			
	Huynh–Feldt	116.775	17.559	6.650			
	Lower-bound	116.775	9.000	12.975			

Bold number indicates that the effect is significant at 5% level.

Let us see how the degrees of freedom are corrected by using the estimates of epsilon suggested by different statisticians in a situation where the sphericity assumption fails. It is very simple, the modified degrees of freedom is obtained by multiplying the value of ε to the degrees of freedom of the treatment (Time) and error (Time) terms in a situation where sphericity is assumed. Thus, the degrees of freedom in the Greenhouse–Geisser correction for the treatment (Time) would be $1.637 (= 0.546 \times 3)$ and that of error would be $14.730 (= 0.546 \times 27)$. Similarly, degrees of freedom can be altered by using the Huynh–Feldt correction. It can be seen in Table 4.5 that the *F* value does not change by applying any of the three corrections. It is so because *F* value is calculated by dividing MSS_{Time} by MSS_{Error}. If the degrees of freedom is altered for the two components, it affects their mean sum of squares as well in a similar fashion, resulting the ratio remains same irrespective of whether the degrees of freedoms are altered or not. This fact can be made clear by understanding the below mentioned formulas:

If sphericity is assumed,

$$F = \frac{\frac{SS_{Time}}{(r-1)}}{\frac{SS_{Error}}{(r-1)(n-1)}} \tag{4.1}$$

where $r (= 4)$ is the number of treatment levels and $n (= 10)$ is the number of subjects. The $(r - 1)$ is the degrees of freedom for the treatment (Time) and $(r - 1)(n - 1)$ is the degrees of freedom for error. For details refer to the Chapter 2.

If sphericity exists, the modified degrees of freedom is obtained by multiplying them with the value of ε

$$F = \frac{\frac{SS_{Time}}{\varepsilon \times (r-1)}}{\frac{SS_{Error}}{\varepsilon \times (r-1)(n-1)}} = \frac{\frac{SS_{Time}}{(r-1)}}{\frac{SS_{Error}}{(r-1)(n-1)}} \tag{4.2}$$

The value of F calculated in equation (4.1) as well as equation (4.2) is same; hence, the value of F remains same irrespective of the fact that whether the sphericity exists or not.

To conclude, if the assumption of sphericity is not violated, that is, Mauchly's test is not significant then no correction is required and the significance value (p) associated with the F for the **Sphericity Assumed** case is reported in the findings. On the other hand, if the sphericity violates then the p value associated with the F for the **Greenhouse–Geisser** case is used for reporting provided the value of estimate of ε given by the Greenhouse–Geisser is less than 0.75. However, if this value of ε is more than 0.75 report the p value associated with the F for the **Huynh–Feldt** case. Since the effect of violating the sphericity increases the p value due to which the results may be entirely different at times if this assumption is not tested in repeated measures design.

Since in the illustration the value of ε is less than 0.75, the Greenhouse–Geisser correction has been used and the p value of 0.009 shall be reported in the findings.

Testing Significance of Within-Subjects Effect   After applying the Greenhouse–Geisser correction, the F value is significant because associated p value of F is 0.009 which is less than 0.05 as shown by the bold face in Table 4.5. In fact, the F value is significant in all the situations and no difference of findings occur due to violation of the sphericity assumption. Since the partial Eta Square is 0.447 which is considered to be moderate, the effect of time is meaningful to enhance the reasoning ability in the meditation intervention program.

Pairwise Comparison of Marginal Means   In repeated measures design, post hoc test cannot be used and therefore there is no option available for such test in SPSS. If F value is found to be significant, the paired t tests are applied to compare each pair of the group means. Due to multiple comparisons of means, the familywise error rate 0.05 in this illustration shall be inflated. To compensate this error, the Bonferroni correction has been used. It automatically adjusts p value for testing the significance of F statistic.

Significance of difference between group means shall be tested as usual at the 0.05 level because the Bonferroni correction enhances the p value accordingly. It can be seen from Table 4.6 that the difference between the group means of zeroth day and 6th week as well as between 2nd week and 6th week are significance because the significance value associated with these two t values are 0.000 and 0.001, respectively, which are less than that of 0.05. However, no difference is found between the means of zeroth day and 2nd week, between zeroth day and 4th week and that of between 2nd week and 4th week.

Marginal Means Plot   The marginal means of reasoning ability measured in different time periods can be shown graphically in Figure 4.10. This output is generated by the SPSS. Additional information regarding the significance level

ILLUSTRATION   87

Table 4.6   Pairwise Comparison of Marginal Means

Measure: Reasoning_ability

		Mean Diff.			95% CI for Difference[a]	
(I) Time	(J) Time	(I − J)	Std. error	Sig.[a]	Lower bound	Upper bound
Zero_day	Week_two	0.100	0.875	1.000	−1.879	2.079
	Week_four	−1.500	1.267	1.000	−4.366	1.366
	Week_six	−3.700*	0.367	**0.000**	−4.529	−2.871
Week_two	Zero_day	−0.100	0.875	1.000	−2.079	1.879
	Week_four	−1.600	1.013	0.893	−3.892	0.692
	Week_six	−3.800*	0.573	**0.001**	−5.097	−2.503
Week_four	Zero_day	1.500	1.267	1.000	−1.366	4.366
	Week_two	1.600	1.013	0.893	−0.692	3.892
	Week_six	−2.200	1.153	0.532	−4.808	0.408
Week_six	Zero_day	3.700*	0.367	**0.000**	2.871	4.529
	Week_two	3.800*	0.573	**0.001**	2.503	5.097
	Week_four	2.200	1.153	0.532	−0.408	4.808

Based on estimated marginal means
[a] Adjustment for multiple comparisons: Bonferroni
*The mean difference is significant at the 0.05 level.
Bold number indicates that the effect is significant at 5% level.

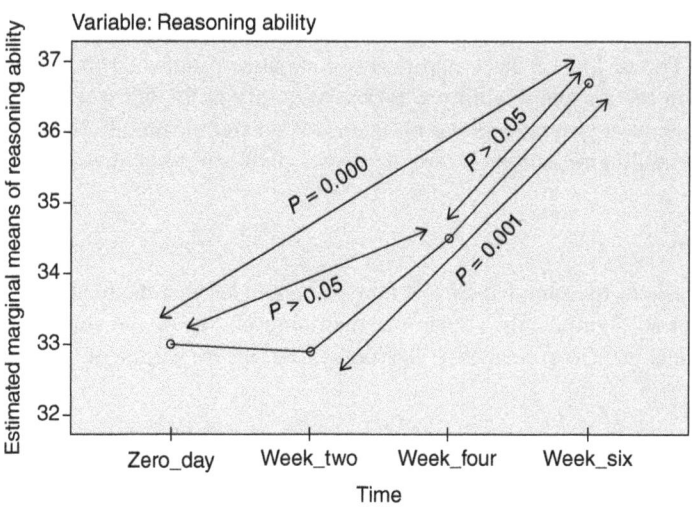

Figure 4.10   Marginal means plot

(as shown in Table 4.6) between the two group means may be added in this figure. It can be seen from Figure 4.10 that the reasoning ability increases in general with the passes of time during meditation intervention program. However, there was no significant increase in reasoning ability till 4th week of the intervention because the p value between zeroth day and 4th week is greater than 0.05 but significant increase was observed in the 6th week because the p value between zeroth day and 6th week is 0.000. Similarly, the p value between the 2nd week and 6th week is also significant because the p value for this difference is 0.001. Thus, based on the sampled data it may be concluded from this figure that the significant increase in the reasoning ability is observed only in 6th week and therefore the meditation program should be implemented at least for 6 weeks to get the significant increase in the reasoning ability.

How to Report the Findings

In reporting the findings in single factor repeated measures design Tables 4.3–4.6 and Figure 4.10 should normally be mentioned. In this illustration the reporting may be made as follows:

Since Mauchly's test was significant (Table 4.4), the sphericity assumption was violated. Since the Greenhouse–Geisser estimate of epsilon (ε) was 0.546 which is less than 0.75, this estimate was used to correct the degrees of freedom. After the correction, the degrees of freedom for finding the significance value of F became (1.637, 14.730) instead of (3, 27). From Table 4.5, it can be seen that after applying the Greenhouse–Geisser correction the F value is significant ($p = 0.009$).

In repeated measures design, no post hoc comparison exists due to which group means were compared using the paired t test. Due to multiple comparisons the familywise error rate inflates. To compensate this error, the Bonferroni correction was applied. The results of the group means comparison indicated that the significant increase in the reasoning ability was observed only in the 6th week of the meditation intervention program. However, there was no significant difference observed in reasoning ability measured on zero_day, week_two, and week_four.

Inference

On the basis of the sampled data, it may be concluded that the meditation intervention program significantly affects the reasoning ability of the subjects. However, the significant effect was observed only after the 6 weeks of the intervention program.

EXERCISE

4.1. Explain situations in which one-way repeated measures design should be used.

4.2. Highlight the benefits of using single factor repeated measures design.

4.3. Discuss design layout of one-way repeated measures design and describe splitting sum of squares into various components along with degrees of freedom.

4.4. Discuss briefly the procedure of solving one-way repeated measures design.

4.5. In a repeated measures design, nine subjects undergo three different treatments T_1, T_2, and T_3. Show the layout design.

4.6. What is the difference between hypothesis construction in case of independent measures design and repeated measures design?

4.7. What is the Bonferroni correction? In comparing five group means if the familywise error rate is taken as 0.05 for testing the significance of F, at what level of significance pair means will be compared and why?

ASSIGNMENT

4.1. A marketing research manager wanted to study the comparative effect of three different types of incentives offered to his dealers on the sale of pizza. He had organized a repeated measures design in which three different types of incentives were offered to six dealers in a random fashion. Each dealer was given all three different types of incentives to see its impact on sale. Following are the sales figure. Apply one-way repeated measures design to investigate the best incentive for improving the sales. Test your hypothesis at the significance level 0.05. Do the following also:

  a. Describe splitting total sum of squares into various components along with their degrees of freedom.

  b. Test normality and sphericity of the data.

Sales Data on Pizza			
	Incentives		
Dealer	Low	Medium	High
1	22	52	50
2	27	65	54
3	21	62	65
4	29	56	45
5	32	68	19
6	35	61	38

4.2. An athletic coach was interested to see the performance trend in improving muscular strength of his athletes during a circuit training program. He organized an experiment in which six athletes participated in the study. They were tested repeatedly for their muscular strength before starting the training and after 2, 4, and 6 weeks of circuit training program. The data so obtained are as follows:

Data on Muscular Strength				
	Time			
Subject	Zero week	2 Week	4 Week	6 Week
1	60	64	72	74
2	65	63	69	75
3	72	73	85	87
4	63	67	74	76
5	58	62	71	73
6	55	57	75	78

Apply repeated measures design to investigate as to which duration is optimum in improving the muscular strength. Test your hypothesis at the significance level 0.05. Also, test the assumptions of normality and sphericity.

BIBLIOGRAPHY

Bakeman. Recommended effect size statistics for repeated measures designs. Behav Res Methods 2005;37(3):379–384. DOI: 10.3758/bf03192707.

Fitzmaurice G, Davidian M, Verbeke G, Molenberghs G, editors. *Longitudinal Data Analysis.* Boca Raton, FL: Chapman and Hall/CRC; 2008. ISBN: 1-58488-658-7.

Jones B, Kenward MG. *Design and Analysis of Cross-Over Trials.* 2nd ed. London: Chapman and Hall; 2003.

Kshirsagar AM, Smith WB. *Growth Curves. Statistics: Textbooks and Monographs 145.* New York: Marcel Dekker, Inc; 1995. ISBN: 0-8247-9341-2.

Pan J-X, Fang K-T. *Growth Curve Models and Statistical Diagnostics. Springer Series in Statistics.* New York: Springer-Verlag; 2002. ISBN: 0-387-95053-2.

Seber GAF, Wild CJ. *Growth Models (Chapter 7). Nonlinear Regression. Wiley Series in Probability and Mathematical Statistics: Probability and Mathematical Statistics.* New York: John Wiley & Sons, Inc; 1989. p 325–367. ISBN: 0-471-61760-1.

Vonesh EF, Chinchilli VG. *Linear and Nonlinear Models for the Analysis of Repeated Measurements.* London: Chapman and Hall; 1997.

5

TWO-WAY REPEATED MEASURES DESIGN

INTRODUCTION

Repeated measures design with two-way ANOVA is used in a situation where the effect of two factors on a dependent variable needs to be investigated simultaneously and both the factors are within subjects. In this design, all subjects are tested in each level of both the factors. This design is also known as within–within design, or, two-way repeated measures design, or, two-way ANOVA with repeated measures. In this design, mean differences between groups that have been split on two within-subject factors are compared.

If in a two-way repeated measures design the factor A has two levels and the factor B has three then there would be six treatment conditions namely A_1B_1, A_1B_2, A_1B_3, A_2B_1, A_2B_2, and A_2B_3. A randomly drawn sample is then tested in all the six treatment conditions. This design is generally used in a situation where individual variations of the subjects cannot be controlled and recruiting large sample for the study is difficult. By using this design a researcher is mainly interested in investigating the following three research questions:

a. Whether factor A affects dependent variable?
b. Whether factor B affects dependent variable?
c. Whether interaction between factors A and B is significant?

The first two research questions are investigated by testing the significance of main effects of A and B. Whereas, the third research question is investigated by

Repeated Measures Design for Empirical Researchers, First Edition. J. P. Verma.
© 2016 John Wiley & Sons, Inc. Published 2016 by John Wiley & Sons, Inc.

investigating the simple effects of A and B. Testing main effects are meaningful only when the interaction effect is not significant. But if the interaction effect is significant then the main effects become meaningless and in that case simple effects of A and B are investigated. The simple effect of factor A refers to comparing the effect of different levels of the factor A on the dependent variable in each level of the factor B. Similarly, the simple effect of the factor B refers to comparing the effect of different levels of the factor B on the dependent variable in each level of the factor A. In two-way repeated measures design, the primary focus of the researcher is to investigate whether an interaction effect between the two factors on the dependent variable exists or not.

It is interesting to note that out of the two independent factors one can be the time duration. For instance, if the researcher's interest is to see the effect of cardio exercise (normal breathing and yogic breathing) as well as time (zero, two, and four weeks) duration on the lungs capacity then the same subjects are tested after performing each exercise at three different time points. Here, cardio exercise and time are the two independent factors.

ADVANTAGES OF USING TWO-WAY REPEATED MEASURES DESIGN

One of the main advantages of the two-way repeated measures design is that the researcher requires limited number of subjects in the experiment. This design allows researchers to complete their experiment more quickly, as fewer subjects are required to complete an entire experiment. Since subjects serve their own control the error variance is reduced resulting increased efficiency in the design. This design can be used for the longitudinal studies where the effect of time needs to be investigated.

ASSUMPTIONS

In using two-way repeated measures design certain assumptions are required to be met. Violation of these assumptions results in inaccurate findings. One of the main assumptions in using the repeated measures design is that the subjects in the study should be randomly selected from the population of interest. Besides this following assumption must holds true for using the design:

a. *Data type* Both the independent factors should be categorical, whereas the dependent variable should be measured on metric scale.

b. *Independence of observations* The measurement obtained on each subject must be independent to each other.

c. *Normality* The observations on dependent variable obtained on the subjects in each treatment combination must be normally distributed. This assumption can be tested by using the Shapiro–Wilk test in SPSS.

d. *Sphericity* Sphericity assumption states that the difference scores computed between any two levels of a within-subjects factor must have the same variance. The sphericity assumption is tested only in a situation where the independent factor has more than two levels. In SPSS, the sphericity assumption is tested by means of Mauchly's *W* test. If this assumption is violated, a correction is made in the degrees of freedom of different sums of squares by means of either the Greenhouse–Geisser or Huynh–Feldt correction.

Testing of the above-mentioned assumptions in using two-way repeated measures design has been discussed in detail while solving the design with SPSS in the illustration later in this chapter.

LAYOUT DESIGN

In two-way repeated measures design both the factors are within-subjects. The layout of this design depends upon whether the levels of the within-subjects factor are different treatment conditions or time period. The layout of the two-way repeated measures design in these two situations has been discussed in the following sections.

Case I: When Levels of Within-Subjects Variable are Different Treatments

In this design all subjects are tested under all treatments. Sufficient time gap is kept between testing of subjects in different levels of each factor in order to ensure that there is no learning or fatigue effect due to testing in the earlier treatment(s). In this design there is an issue of order effect. This effect includes practice, boredom, and fatigue. Generally, performance of subjects either improves or deteriorates. Repeated measures designs are almost always affected by the order effect. However, the only exception for this effect is seen in longitudinal studies, that is, when one of the independent factors is a time.

Counterbalancing is used to counteract order effect. This is done by dividing sample into *k* groups and randomizing treatments in each group of samples independently where *k* is the number of levels of within-subjects variable. Let us suppose that within-subjects factor has three levels then the first group will undergo the first treatment condition thereafter second and then the third one, whereas, the second group may undergo into the second treatment condition first and then the second and later the third in the sequence. Similarly, order of the third group may be the third treatment conditions first then the second and thereafter the first.

Let us consider an experiment where two-way repeated measures design is used to investigate the effect of caffeine (coffee and placebo) and time of testing on the mathematical ability of subjects. Here, caffeine and time are the two within-subjects factors. If the sample consists of six randomly selected subjects then the layout of the two-way repeated measures design can be shown by Figure 5.1. As per the testing protocol, on the first day the subjects S1 and S2 are tested for their mathematical ability test in the morning, the subjects S3 and S4 are tested in the afternoon, and the subjects S5 and S6 get tested in the evening. Similarly, the testing protocol for the subjects in placebo group can be done by randomizing treatments to the subjects. All subjects are tested after consuming coffee. The only care which the researcher must take is to ensure that all the subjects should be tested in each treatment condition.

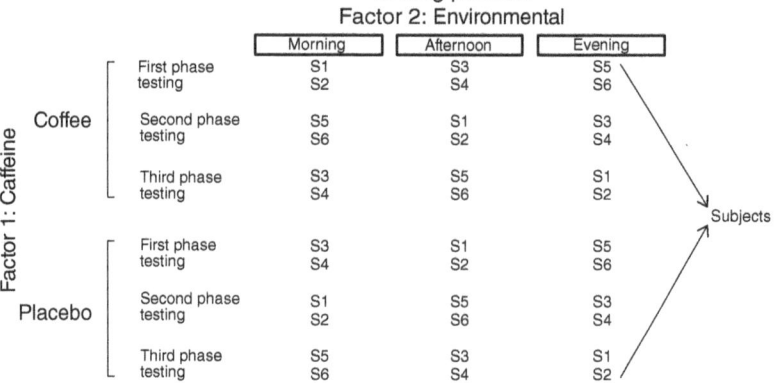

Figure 5.1 Layout of the two-way repeated measures design.

Case II: When the Levels of the Within-Subjects Variable are Different Time Durations

In this case, all subjects in each level of the first factor are repeatedly tested at different time points during treatment. Here performance of the subjects, while testing in a particular point is affected due to the cumulative effect of treatment in the earlier durations, the issue of counter balancing and order effect is irrelevant because the purpose of the study is to see the effectiveness of treatment in different time duration. Let us consider that the above-mentioned experiment is organized to see the effect of caffeine on mathematical ability in four different time duration, that is, before experiment, after three, six, and nine weeks. If sample consists of six randomly selected subjects then the layout of this two-way repeated measures design can be shown by Figure 5.2.

APPLICATION

Two-way repeated measures design has many applications in the area of psychology, management, social sciences, education, medicine and sports. Some of the applications of this design is listed below:

1. A psychologist may wish to investigate the effect of caffeine on memory retention over a period of time. To do so an experiment may be planned in which the subjects may be tested for their memory power before experiment, after one and two weeks of caffeine consumption. Further, after some time the same subjects may be tested at these time points when they are under placebo treatment.
2. A researcher may be interested to see the impact of fat consumption and time of the day on the performance in a comprehension test. A random sample drawn from the population of interest may be tested for their performance in the comprehension test during morning, afternoon, and evening while being fed with

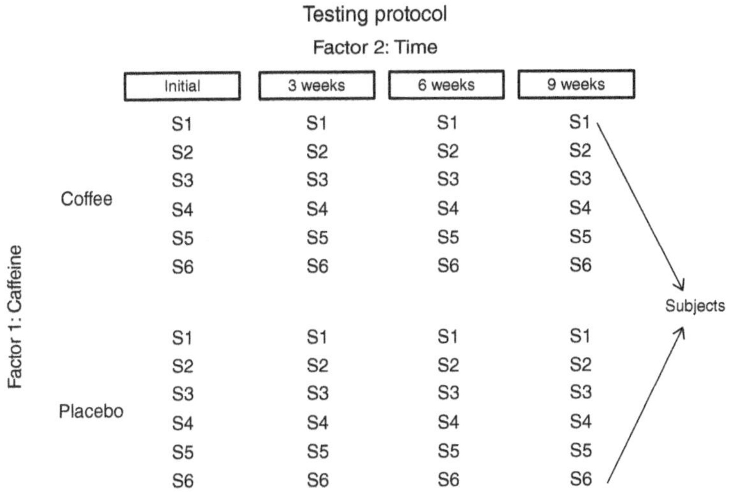

Testing protocol

Factor 2: Time

	Initial	3 weeks	6 weeks	9 weeks

Coffee

S1	S1	S1	S1
S2	S2	S2	S2
S3	S3	S3	S3
S4	S4	S4	S4
S5	S5	S5	S5
S6	S6	S6	S6

Subjects

Placebo

S1	S1	S1	S1
S2	S2	S2	S2
S3	S3	S3	S3
S4	S4	S4	S4
S5	S5	S5	S5
S6	S6	S6	S6

Factor 1: Caffeine

Figure 5.2 Layout of the two-way repeated measures design when one of the factors is time.

no fat, medium fat, and high-fat diet. Thus, all the subjects in the sample are tested for their performance in all the nine treatment conditions. The main interest of the researcher in using two-way repeated measures design is to test the interaction effect between fat and time of testing.

3. A market researcher may like to investigate the effect of time and season on sale in grocery outlets of a company. Here, similar level of grocery outlets may be randomly selected by the researcher and their sales figure may be recorded in both the shifts (first half and second half of the day) in five different months equally spread during the year. The shifts and the time are the two independent variables whose effects may be investigated on the sales figure.

STEPS IN SOLVING TWO-WAY REPEATED MEASURES DESIGN

In two-way repeated measures design, a researcher's interest is to study the effect of two within-subjects factors simultaneously on some dependent variable. The main focus of the investigator is to see whether interaction effect between two factors is significant or not.

Following steps are used in solving the two-way repeated measures design:

1. Check normality assumption for the data in all treatment conditions by applying the Shapiro–Wilk test in SPSS (for details refer to Chapter 3)
2. Describe layout of the design
3. Write research questions which need to be answered
4. Formulate hypotheses to be tested
5. Decide familywise level of significance (α) at which the hypothesis is required to be tested

6. Use SPSS commands to generate the following outputs
   a. Mauchly's test of sphericity
   b. Descriptive statistics
   c. F-table for testing within-subjects effects
   d. Estimates of marginal means of independent variables if F value for the independent variables are significant
   e. Pairwise comparison of marginal means of variables if F value for the independent variables are significant
   f. Marginal mean plots of independent variables if F values for the variables are significant
   g. Marginal mean plots for interaction if F value for the interaction is significant

7. If interaction is significant perform the following repeated measures ANOVAs by using the same data file in SPSS to test the simple effects.
   a. Repeated measures ANOVAs for testing the simple effect of the first factor (within-subjects) and generate the following outputs:
      i. Mauchly's test of sphericity in different levels of the second factor (within-subjects).
      ii. F-table for testing significance of the first factor effect in each level of the second factor.
      iii. Pairwise comparisons of means in each level of the second factor.
   b. Repeated measures ANOVAs for testing the simple effect of the second factor (within-subjects) and generate the following outputs:
      i. Mauchly's test of sphericity in different levels of the first factor (within-subjects).
      ii. F-table for testing significance of the second factor effects in each level of the first factor.
      iii. Pairwise comparisons of means in each level of the first factor.

8. Test sphericity assumption for each factor and the interaction by means of Mauchly's test. If Mauchly's test is significant ($p < 0.05$) for any effect, the sphericity assumption is said to be violated and in that case apply correction in the degrees of freedom for testing the significance of F value.

9. In case sphericity is violated for an effect, look for the value of epsilon (ε). If its value is less than 0.75, apply the Greenhouse–Geisser correction otherwise apply the Huynh–Feldt correction and test the significance of F value by looking to the p value attached with the F.

10. If F ratio is significant do the pairwise comparison among group means by applying the Bonferroni correction.

11. Report the findings.

The procedure involved in solving this design shall be explained by means of the below-mentioned illustration.

ILLUSTRATION 97

ILLUSTRATION

An experiment was conducted to investigate the effect of environment and music on the employee's performance in a cottage industry of packaging. Six subjects were randomly selected amongst those who were engaged in packaging job of matchbox. The investigator created three environments (hot, humid, and cold) in a laboratory and selected three types of music settings (Instrumental, Jazz, and No music) for the study. Thus, there were nine treatment conditions (environment × music). All six subject's performance was tested under each of the nine treatment conditions and the number of matchbox prepared in a day was recorded which is shown in the Table 5.1. Let us see how two-way repeated measures ANOVA can be used to draw the inferences by using the SPSS software.

Layout Design

In this illustration, the effect of environment and music is to be seen on the packaging task. Since all six subjects were tested in all the nine treatment conditions (environment × music) both the independent variables, that is, environment and music are within-subject factors. Thus, design in this illustration is a within–within design, layout of which can be shown by Figure 5.3.

Table 5.1 Number of Match Box Prepared Per Hour in a Day

	Environment		
	Hot	Humid	Cold
Music	No music		
	20	16	27
	18	17	24
	22	16	26
	16	19	17
	18	20	26
	20	22	23
	Jazz		
	22	21	23
	20	25	21
	24	27	22
	19	21	20
	22	27	25
	20	26	25
	Instrumental		
	24	26	21
	26	22	20
	25	22	18
	26	21	24
	24	19	18
	25	22	21

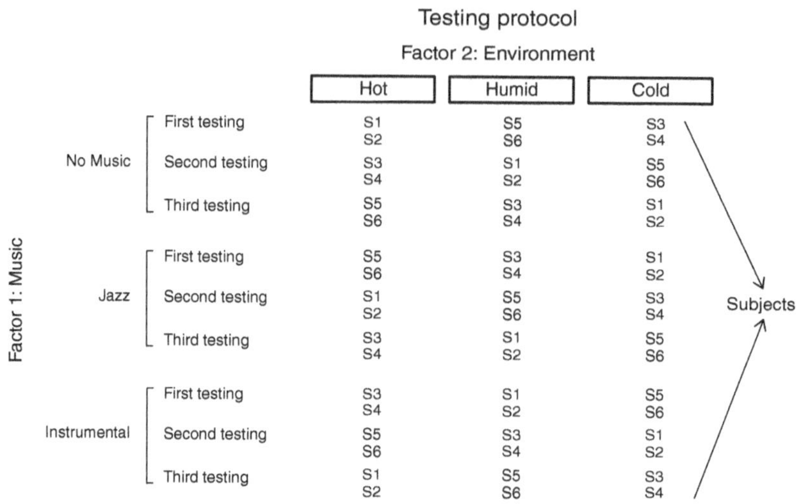

Testing protocol

Factor 2: Environment

			Hot	Humid	Cold

Figure 5.3   Layout of the repeated measures design with two factors

In this design, all the subjects undergo each of the nine treatment conditions but not in a particular sequence. It can be seen from the testing protocol shown in Figure 5.3 that on the first day of testing two subjects, S1 and S2 are exposed to the no music × hot, S3 and S4 to no music × cold and S5 and S6 to no music × humid treatment conditions. This order has been randomized on the second and third day of testing under different environment conditions when there is no background music. Similarly, subjects have been randomized with background music of Jazz and Instrumental as well. The readers can randomize the allocation of treatment conditions to the subjects in a different manner also. The only care one should take is that in each of the nine treatment conditions all subjects should be exposed. The purpose of randomizing the sequence is to done away the order effect in the performance of the subjects due to specific order of testing. This method of randomizing the allocation of treatments to the subjects is known as counterbalancing.

Distribution of Variation and Degrees of Freedom

To understand two-way repeated measure design, it is necessary to understand as to how the total variability is distributed. Figure 5.4 shows the breakup of total variability into different components along with their associated degrees of freedom in a within–within design. Here, r and c are the levels of the first and second independent factors (within-subjects) and n is the number of subjects in the experiment. Since data in Table 5.1 has been obtained in the experiment by using the within–within design where all the $n(= 6)$ subject's performance was tested under all the $rc(3 \times 3 = 9)$ treatment conditions, the total sum of squares (TSS) has been first partitioned into two components: sum of squares between subjects ($SS_{subjects}$) and sum of squares within subjects ($SS_{Within\_subjects}$). In this design, the between subjects sum of squares

ILLUSTRATION   99

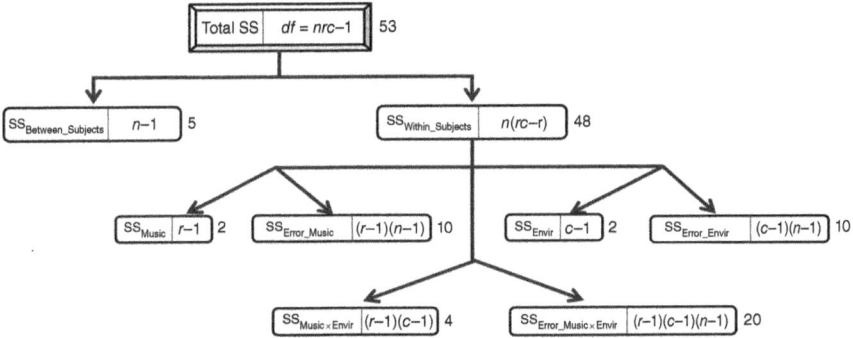

Figure 5.4 Scheme of distributing total sum of squares and degrees of freedom in the two-way repeated measures design

is separated from the total variation to eliminate the variation due to individual differences thus resulting into lesser experimental error. The between sum of squares is not partitioned further.

The within subjects sum of squares is further partitioned into six components: Music sum of squares (SS_{Music}), Within music sum of squares (SS_{Error_Music}), Environment sum of squares (SS_{Envir}), Within environment sum of squares (SS_{Error_Envir}), Music $\times$ Environment sum of squares ($SS_{Music\times Envir}$) and Within music $\times$ environment sum of squares ($SS_{Error\_Music\times Envir}$). The Music sum of squares is the variation among the means of the r music groups averaged over all the levels of the Environment. The sum of squares within music is the variation due to the joint effect of the Music and subjects within group and is used as an error term to test the F ratio of Music. The Environment sum of squares is the variation among the means of the c environment groups averaged over all the levels of the Music. The sum of squares within environment is the variation due to the joint effect of the Environment and subjects within group and is used as an error term to test the F ratio of Environment. The sum of squares due to Music $\times$ Environment is the variation due to the joint effect of music and environment. The sum of squares within Music $\times$ Environment is the variation due to the joint effect of music, environment, and subjects within groups and is used as an error term to test the F ratio of Music $\times$ Environment interaction. Thus, the total sum of squares in the two-way repeated measures design can be broken in different components as shown in Figure 5.4.

$$\text{Total SS} = SS_{Subjects} + SS_{Within\_Subjects}$$

$$= SS_{Subjects} + (SS_{Music} + SS_{Error\_Music}) + (SS_{Envir} + SS_{Error\_Envir})$$

$$+ (SS_{Music\times Envir} + SS_{Error\_Music\times Envir})$$

In this illustration of two-way repeated measures design, the total degrees of freedom, that is, $53(= nrc - 1)$ has been partitioned into $5(= n - 1)df$ due to variation between subjects and $48(= n(rc - 1))df$ due to variation within subjects. Further, $48(= n(rc - 1))df$ of within-subjects variation has been partitioned into six components namely, $2(= r - 1)df$ for Music, $10[= (r - 1)(n - 1)]df$ for Error_Music, $2(= c - 1)df$ for Envir, $10[= (c - 1)(n - 1)]df$ for Error_Envir, $4[= (r - 1)(= c - 1)]df$ for Music $\times$ Envir and $20[= (r - 1)(c - 1)(n - 1)]df$ for Error_Music $\times$ Envir.

Research Questions

The following three research questions need to be investigated in this illustration:

a. Whether background music affects worker's performance..
b. Whether performance of workers is affected by the environment.
c. Whether interaction between background music and type of environment affects worker's performance. In other words, whether pattern of packaging performance under three environmental conditions differs in each of the three background music. And similarly whether performance pattern of workers under different background music differs in each environment condition.

Hypotheses Construction

To investigate the above-mentioned research questions following hypotheses shall be tested:

a. *Main effect of Music*

$$H_0 : \mu_{\text{No\_Music}} = \mu_{\text{Jazz}} = \mu_{\text{Instrumental}}$$

against H_1: At least one group mean differs
b. *Main effect of Environment*

$$H_0 : \mu_{\text{Hot}} = \mu_{\text{Humid}} = \mu_{\text{Cold}}$$

against H_1: At least any one group mean differs
c. *Interaction effect (Music $\times$ Environment)*
H_0: There is no interaction between Music and Environment
against
H_1: The interaction between Music and Environment is significant

Testing the first two hypotheses shall answer the first two research questions (main effects) mentioned above, whereas the third research question can be investigated by testing the significance of the interaction effect. One should understand that testing main effects becomes meaningful only when the interaction is not significant. If the F ratio for the main effect is significant then it should be further investigated by using the pairwise comparison of means only in a situation where the interaction effect is not significant. If the interaction effect is significant, then the whole focus of the experiment is to test the simple effects only. Simple effect refers to comparing packaging performance of three music categories in each level of the environment and comparing subject's performance under three different types of environments in each music category.

ILLUSTRATION

101

Level of Significance

In this illustration, familywise error rate (α) shall be taken as 0.05. This is a within–within design where both the factors are repeated measures. Since post hoc test for the repeated measures design does not exist hence if the F value of any of the factors, that is, Music or Environment is significant then the Bonferroni correction shall be applied for correcting the level of significance. In repeated measures design, no single error sum of squares is computed unlike independent measures design; hence, simple effects need to be evaluated separately by applying three one-way repeated measures ANOVA for each within-subjects variable. This will inflate the familywise error rate (α). To compensate this, appropriate correction in the level of significance would be made while testing the significance of F in testing the simple effect.

Solving Repeated Measures Design with Two-Way ANOVA Using SPSS

The procedure involved in solving the two-way repeated measures design shall be discussed by using the SPSS software. The first thing is to prepare a data file in SPSS. In **Variable View** define all the following nine treatment combinations as variables.

NOM_Hot

NOM_Humid

NOM_Cold

Jz_Hot

Jz_Humid

JZ_Cold

Inst_Hot

Inst_Humid

Inst_Cold

The data needs to be entered column-wise as shown in Figure 5.5. After preparing the data file, follow the sequence of commands mentioned below to get the screen shown in Figure 5.6 for defining independent and dependent variables.

Analyze → General Linear Model → Repeated Measures

Factor 1 is written in "Within-Subject Factor Name" section by default. Replace this by the first independent variable *Music* and write the number of levels as 3 as there are three levels (no music, jazz, and instrumental). Click on **Add** to enter this information into the box. Write the name of second independent variable *Environment* in the "Within-Subject Factor" section and write the number of levels as 3 because *Environment* has also three levels (hot, humid, and cold). Click on **Add** to enter this information into the box. In the "Measure Name" section type *Packaging* as a dependent variable. Click on **Add** to enter this information into the box. Readers' should note that the first letter of the any variable should always be alphabet followed by any letter or number. If name consists of two or more words, then they should be joined by the underscore.

	NOM_Hot	NOM_Humid	NOM_Cold	Jz_Hot	Jz_Humid	JZ_Cold	Inst_Hot	Inst_Humid	Inst_Cold
1	20.00	16.00	27.00	22.00	21.00	23.00	24.00	26.00	21.00
2	18.00	17.00	24.00	20.00	25.00	21.00	26.00	22.00	20.00
3	22.00	16.00	26.00	24.00	27.00	22.00	25.00	22.00	18.00
4	16.00	19.00	17.00	19.00	21.00	20.00	26.00	21.00	24.00
5	18.00	20.00	26.00	22.00	27.00	25.00	24.00	19.00	18.00
6	20.00	22.00	23.00	20.00	26.00	25.00	25.00	22.00	21.00

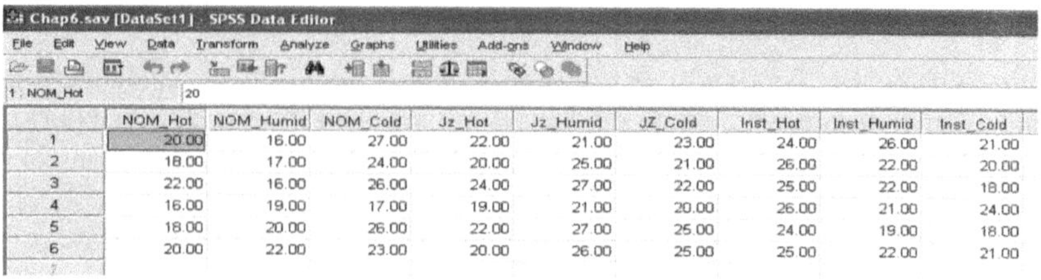

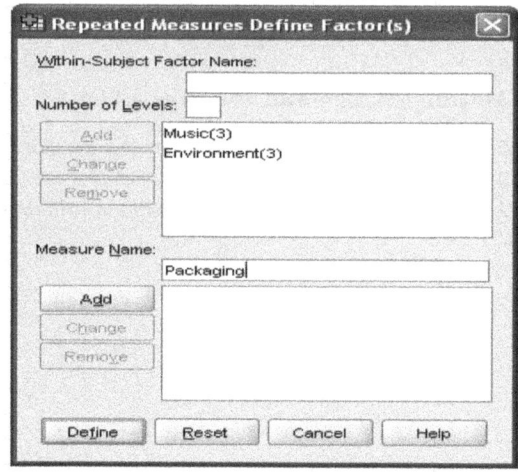

Figure 5.5 Data format in the repeated measures design with two factors

Figure 5.6 Screen showing options for defining independent and dependent variables

Clicking on the command **Define** in the screen shown in Figure 5.6 shall take you to the screen as shown in Figure 5.7 for selecting levels of the Within-subjects factors. Select all nine variables from the left panel and bring them into the "Within-Subjects Variables" section of the screen in the right panel.

After selecting variables, option needs to be defined for generating various means plot in the outputs of this design. Click on the **Plots** command to get the screen as shown in Figure 5.8. Do the following:

a. Select the variable *Environment* from the "Factors" section and bring it to the "Horizontal Axis" area for generating means plot of the main effect of Environment on packaging. This plot is used to compare the main effect of within-subjects factor (Environment) on packaging. Click on **Add.**
b. Select the variable *Music* from the "Factors" section and bring it to the "Horizontal Axis" area for generating means plot of the main effect of Music on packaging. This plot is used to compare the main effect of the within-subjects factor (Music) on packaging. Click on **Add.**

ILLUSTRATION   103

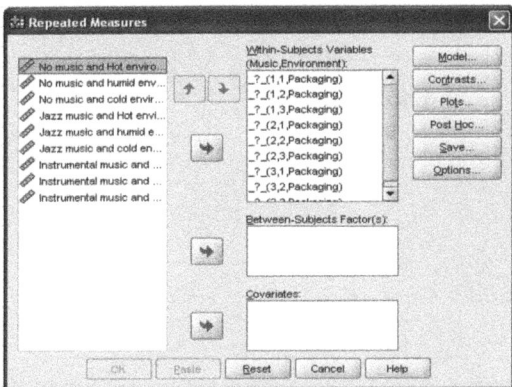

Figure 5.7   Screen showing option for selecting variables defining all treatment combinations

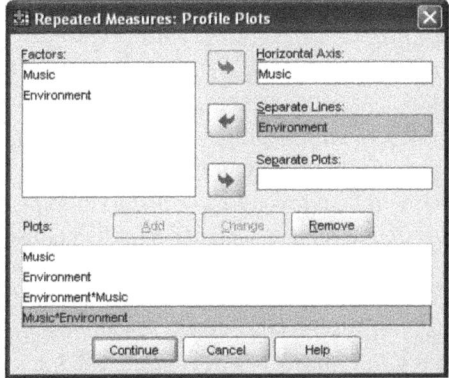

Figure 5.8   Screen showing option for means plot

c. For generating the means plot of interaction, bring the variables Environment and Music from the "Factors" section to the "Separate Lines" and "Horizontal Axis" sections, respectively. This plot is helpful in comparing simple effect of music in each level of the environment. Click on **Add** to get the means plot in the output.

d. Similarly for comparing the simple effects of environment in each level of the music, bring the variables Music and Environment from the "Factors" section to the "Separate Lines" and "Horizontal Axis" sections, respectively. Click on **Add** and **Continue.**

This will take you back to the screen as shown in Figure 5.7. Click on **Options** command to get the screen as shown in Figure 5.9 for selecting options to generate all required outputs in the analysis.

Transfer the variables *Environment, Music* and *Environment*Music* from the "Factor(s) and Factor Interactions" section into the "Display Means for" section. This will

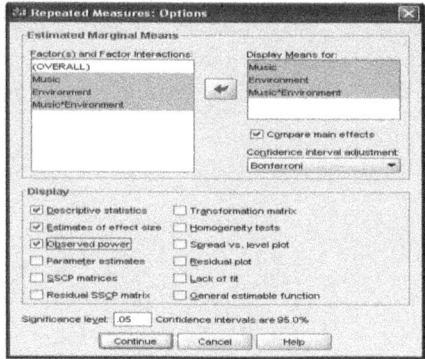

Figure 5.9 Screen showing option for computing various outputs in the repeated measures design with two factors

generate marginal means of Environment, Music, and means within each combination of Environment × Music.

Check the option 'Compare main effects'. Select 'Bonferroni' correction by clicking on the sign ▼ mentioned on the "LSD(none)" tag. This will generate the output for testing the main effects of within-subjects factors Environment and Music. Since there is no post hoc test available for within-subjects factor, Bonferroni correction is used to compensate the inflated type I error due to multiple comparison of group means. Readers are advised to refer to Chapter 3 for detail discussion.

Check the option 'Descriptive statistics' for computing mean and standard deviation of the data in each treatment condition. Ensure that the level of significance is selected as 0.05. In fact, it is selected by default. Let all other options remain selected as it is. Check the option 'Estimates of effect size'. Click on **Continue** to get back the screen shown in Figure 5.7. Click on **OK** to get the output.

SPSS Output and Interpretation

For analyzing this design, SPSS generates lots of output in the output window, but only relevant outputs as mentioned below need to be selected for interpretation.

1. Mauchly's test of sphericity (Table 5.2)
2. Descriptive statistics (Table 5.3)
3. F-table for testing within-subjects effects (Table 5.4)
4. Estimates of marginal means of Music (Table 5.5)
5. Pairwise comparison of marginal means of Music (Table 5.6)
6. Means plot of Music (Figure 5.10)
7. Means plot of Music × Environment (Figure 5.15)
8. Means plot of Environment × Music (Figure 5.16)

The output in Table 5.4 is used to test the significance of main effects of within-subjects factors: Environment, Music, and Interaction between Environment × Music. In case of significant interaction, investigating main effects becomes

ILLUSTRATION   105

meaningless and in that case simple effects are investigated by using the same data file in SPSS. The procedure has been discussed in the later part of this chapter. Thus, in a situation where interaction is significant, following additional outputs are generated to discuss simple effects of both within-subjects factors:

1. Repeated measures ANOVAs for testing the simple effect of Environment (within-subjects)
   a. Mauchly's test of sphericity in different music categories (Table 5.7)
   b. F-table for testing significance of Environment (within-subjects) effects in each music category (Table 5.8)
   c. Pairwise comparisons of means in each music category (Table 5.9)
2. Repeated measures ANOVAs for testing the simple effect of Music (within-subjects)
   a. Mauchly's test of sphericity in different environment categories (Table 5.10)
   b. F-table for testing significance of Music (within-subjects) effects in each environment category (Table 5.11)
   c. Pairwise comparisons of means in each environment category (Table 5.12)

Testing Assumptions

For using within–within design, certain assumptions need to be satisfied. Let us check these assumptions for the data shown in the illustration first.

Table 5.2   Mauchly's Test of Sphericity

		Measure: Packaging						
						Epsilon		
Within-Subjects Effect	Mauchly's W	Approx. Chi- Square	df	Sig. (p value)	Greenhouse– Geisser	Huynh– Feldt	Lower-bound
Music	0.196	6.527	2	**0.038**[*]	0.554	0.597	0.500
Environment	0.950	0.205	2	0.903	0.953	1.000	0.500
Music*Environment	0.069	9.157	9	0.477	0.456	0.705	0.250

*Significant at 0.05 level.

Table 5.3   Descriptive Statistics

	Mean	SD	N
NOM_Hot	19.0000	2.09762	6
NOM_Humid	18.3333	2.42212	6
NOM_Cold	23.8333	3.65605	6
Jz_Hot	21.1667	1.83485	6
Jz_Humid	24.5000	2.81069	6
JZ_Cold	22.6667	2.06559	6
Inst_Hot	25.0000	0.89443	6
Inst_Humid	22.0000	2.28035	6
Inst_Cold	20.3333	2.25093	6

Data Type   Since both the independent variables, that is, music and environment, are categorical and dependent variable packaging performance is a metric variable, the assumption about the data type holds true.

Independence of Measurement   Since performance of each subject has been tested independently and does not depend upon the performance of the other subjects, the data can be considered to be independent to each other.

Normality   To test normality of all nine data sets in different treatment conditions the Shapiro–Wilk statistic has been shown in Table 5.13. This test has been computed by using the SPSS. For details the readers are advised to refer to the Chapter 3. Since none of the value of the Shapiro–Wilk statistic is significant ($p > 0.05$), the assumption of normality holds true for all the nine data sets in different treatment conditions.

Sphericity   In within–within design sphericity assumption is required to be tested in each within-subjects factor as well as interaction. This assumption refers to the equal variances of the differences between all combinations of within-subjects groups. Alternatively, correlations among the repeated measures are not supposed to differ significantly. The sphericity assumption is tested by using the Mauchly's *W* test generated in the SPSS outputs. The sphericity assumption is not violated if Mauchly's statistic is nonsignificant. It can be seen from Table 5.2 that the Mauchly's *W* test is significant only in Music ($p < 0.05$), whereas it is nonsignificant in Environment ($p > 0.05$) and Interaction both ($p > 0.05$). Thus, the assumption of sphericity is violated only in Music factor but not in Environment and Interaction.

Correction for Sphericity   Since in this illustration sphericity assumption has been violated in Music factor, correction is required in its related degrees of freedom whereas no correction is required for the factor Environment and Interaction. Since Epsilon (ε) value for Music is less than 0.75, the Greenhouse–Geisser correction shall be used for testing the significance of *F* for Music. For detail discussion on sphericity readers may refer to Chapters 2 and 3.

Descriptive Statistics

Table 5.3 indicates mean and standard deviation of each of the nine treatment groups. This content can be used to compare the mean and standard deviation in each cell. These means are used to prepare means plots for comparing main and simple effects.

Testing Main Effect of Music (Within-Subjects)

In this repeated measures design, there are two independent factors Environment and Music whose effect needs to be investigated. Both these factors are repeated measures or within-subjects. In this illustration, interaction effect is significant; hence, analyzing the main effects is not required and the meaningful conclusions can be obtained by analyzing the simple effects only. However, the main effect of Music has been discussed in detail to show the procedure involved in it.

Table 5.4 F-table for Testing the Significance of Within-Subjects Effects

Source		Type III SS	df	Mean Square	F (p value)	Sig.[a]	Partial Eta Squared
		Measure: packaging					
Music	Sphericity Assumed	60.259	2	30.130	3.800	0.059	0.432
	Greenhouse–Geisser	**60.259**	**1.108**	**54.366**	**3.800**	**0.102**	**0.432**
	Huynh–Feldt	60.259	1.195	50.427	3.800	0.097	0.432
	Lower-bound	60.259	1.000	60.259	3.800	0.109	0.432
Error(Music)	Sphericity Assumed	79.296	10	7.930			
	Greenhouse–Geisser	79.296	5.542	14.308			
	Huynh–Feldt	79.296	5.975	13.272			
	Lower-bound	79.296	5.000	15.859			
Environment	**Sphericity Assumed**	4.593	2	2.296	1.030	0.392	0.171
	Greenhouse–Geisser	4.593	1.905	2.411	1.030	0.390	0.171
	Huynh–Feldt	4.593	2.000	2.296	1.030	0.392	0.171
	Lower-bound	4.593	1.000	4.593	1.030	0.357	0.171
Error(Environment)	Sphericity Assumed	22.296	10	2.230			
	Greenhouse–Geisser	22.296	9.525	2.341			
	Huynh–Feldt	22.296	10.000	2.230			
	Lower-bound	22.296	5.000	4.459			
Music*Environment	**Sphericity Assumed**	**204.074**	**4**	**51.019**	**8.768**	**0.000**	**0.637**
	Greenhouse–Geisser	204.074	1.825	111.826	8.768	0.008	0.637
	Huynh–Feldt	204.074	2.819	72.400	8.768	0.002	0.637
	Lower-bound	204.074	1.000	204.074	8.768	0.031	0.637
Error(Music*Environment)	Sphericity Assumed	116.370	20	5.819			
	Greenhouse–Geisser	116.370	9.125	12.753			
	Huynh–Feldt	116.370	14.094	8.257			
	Lower-bound	116.370	5.000	23.274			

[a]Computed using alpha = 0.05

Since sphericity assumption has been violated for Music factor, its F value will be tested for its significance by applying the Greenhouse–Geisser correction. Table 5.4 shows that the F for Music is not significant ($p = 0.102$) because its associated p value is more than 0.05. Thus, the null hypothesis that the average packaging performance scores is same in all the three music groups irrespective of environment categories is not rejected at 5% level. Since the main effect of Music is nonsignificant, no further analysis is required. However, pairwise comparison among different music groups irrespective of the environment has been shown in Table 5.6 in order to make the procedure clear to the readers.

Pairwise Comparison of Marginal Means of Music Groups   Since post hoc test cannot be applied for the repeated measures design, no option for this test is shown in SPSS for comparing the marginal group means of music. In case of repeated factor, SPSS uses paired t-test for comparing each pair of the marginal group means. Due to multiple comparisons the level of significance inflates and to compensate this error option for Bonferroni correction (for detail see Chapter 2) has been used in SPSS (Figure 5.9) while comparing. Table 5.6 generated in the SPSS output provides the pairwise comparison of marginal means of different music groups.

Since F value for the Music is not significant, no conclusion should be drawn about the difference of performance in three different environment groups.

Means Plot of Music   Figure 5.10 shows the means plot of the marginal means of music groups. This means plot has been generated by using the marginal means of the environment group shown in Table 5.5. It can be noted from this figure that the subjects performance was higher in Jazz group in comparison to that of No music and Instrumental group irrespective of the environment. However, significance of the difference in their performance cannot be established as the F ratio is not significant.

Testing Main Effect of Environment (Within-Subjects)

Table 5.4 shows that the F value for Environment is not significant ($p = 0.392$); hence, it may be concluded that the main effect of environment is not significant and the null hypothesis may not be rejected at 5% level. Thus, it may be inferred that the packaging performance of the subjects do not differ in three environment groups irrespective of the music backgrounds. Since F for Environment is not significant, no further analysis is required.

Testing Significance of Interaction (Environment × Music)

Table 5.4 reveals that the F value for the interaction (Environment × Music) is significant because the p value associated with the F (Sphericity Assumed) is 0.000 which is less than 0.05. Further, partial Eta Square for the interaction is 0.637 which is considered to be very high; hence, analyzing interaction will be meaningful in the study. In other words, the null hypothesis of no interaction between environment and music is rejected at 5% level. Since the sphericity assumption has not been violated for the

ILLUSTRATION   109

Table 5.5   Estimates of Marginal Means of Music

		Measure: Packaging		
			95% Confidence Interval	
Environment	Mean	Std. Error	Lower Bound	Upper Bound
No Music	20.389	0.674	18.655	22.123
Jazz	22.778	0.724	20.918	24.638
Instrumental	22.444	0.521	21.105	23.784

Table 5.6   Pairwise Comparison of Marginal Means of Music

		Measure: Packaging				
(I) Music	(J) Music	Mean Difference (I-J)	Std. Error	Sig[a]	95% CI for Difference[a]	
					Lower Bound	Upper Bound
No Music	Jazz	−2.389[b]	.338	0.003	−3.583	−1.195
	Instrumental	−2.056	1.049	0.322	−5.761	1.650
Jazz	No Music	2.389[b]	0.338	0.003	1.195	3.583
	Instrumental	0.333	1.196	1.000	−3.892	4.559
Instrumental	No Music	2.056	1.049	0.322	−1.650	5.761
	Jazz	−0.333	1.196	1.000	−4.559	3.892

Based on estimated marginal means
[a] Adjustment for multiple comparisons: Bonferroni.
[b] The mean difference is significant at the 0.05 level.

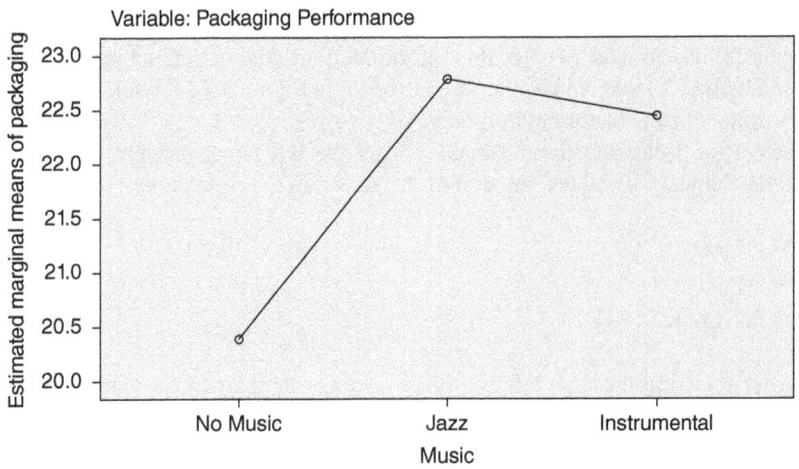

Figure 5.10   Marginal means plot of Music

interaction(Environment * Music), no correction shall be made to the degrees of free-dom and the *p* value against "Sphericity Assumed" row shall be considered for testing the significance of *F*. Since interaction is significant, it is important to investigate the simple effects of Environment as well as Music to have clearer picture.

Type I Error for Simple Effect   In finding simple effect, three repeated measures ANOVA tests shall be applied for each within-subjects variable (Environment and Music). Thus, the level of significance (α) for testing the significance of *F* for Environment in each Music condition shall be 0.017(=0.05/3). Similarly in testing the significance of Music in each Environment condition also the level of significance (α) will be 0.017.

Simple Effect of Environment (Within-Subjects)   Simple effect of Environment can be investigated by comparing the packaging scores of all the three environment groups in each music category separately. In other words, the null hypothesis of no difference between the packaging scores in all the three environment groups needs to be tested separately in each music category. To do so three separate one-way repeated measures ANOVA needs to be applied.

Using the same data file as shown in Figure 5.5, follow the sequence of commands mentioned below to get the screen shown in Figure 5.11 for defining independent and dependent variables.

Analyze → General Linear Model → Repeated Measures

Factor 1 is written in "Within-Subject Factor Name" section by default. Replace this by the first independent variable *Environment_NoMusic* and enter the number of levels as 3 as there are three levels (hot, humid, and cold). This option will facilitate the comparison of the packaging performance of the three environment groups when no background music is provided to all the subjects. Click on **Add** to enter this information into the box. In the "Measure Name" section, type *Packaging* as a dependent variable. Click on **Add** to enter this information into the box. Clicking on the command **Define** will take you to the screen shown in Figure 5.12 for selecting levels of the within-subjects factor environment.

Select the following three variables from the left panel and bring them to the "Within-Subjects Variables" section of the screen in the right panel.

NOM_Hot
NOM_Humid
NOM_Cold

After selecting these variables, option needs to be defined for generating means plot of environment in no music group. Click on the **Plots** command to get the screen as shown in Figure 5.13. Select the factor *Environment_NoMusic* from the "Factors" section and bring it to the "Horizontal Axis" area for generating means plot. This plot is used to compare the simple effect of Environment on packaging in no music group. Click on **Add** and **Continue**.

ILLUSTRATION

111

This will take you back to the screen as shown in Figure 5.12. Click on **Options** command to get the screen as shown in Figure 5.14 for selecting options to generate all required outputs in one-way repeated measures ANOVA.

Bring the factor *Environment_NoMusic* from the "Factor(s) and Factor Interactions" section into the "Display Means for" section. This will generate means of different levels of Environment in no music group.

Check the option 'Compare mean effects'. Select the Bonferroni correction by clicking on the sign ▼ mentioned on the "LSD(none)" tag. This will generate the output for testing the simple effect of Environment (within-subjects) in no music group.

Check the option 'Descriptive statistics' for computing mean and standard deviation of the data in each treatment condition. Ensure that the level of significance is selected as 0.05. In fact, it is selected by default. Let all other options remain as it is. Check the option 'Estimates of effect size' Click on **Continue** to get back to the screen shown in Figure 5.12. Click on **OK** to get the outputs shown in Tables 5.7–5.9.

Remark The outputs in investigating the simple effect of Environment separately in Jazz and Instrumental music groups can be obtained by following the above-mentioned procedure.

1. For comparing the effect of environment in Jazz group define the factor name as Environment_Jazz in Figure 5.11 and select the following three variables in the screen shown in Figure 5.12.

   Jz_Hot

   Jz_Humid

   Jz_Cold

2. Similarly for comparing the effect of environment in Instrumental group define the factor name as Environment_Inst in the screen shown in Figure 5.11 and select the following three variables in the screen shown in Figure 5.12

   Inst_Hot

   Inst_Humid

   Inst_Cold

By combining outputs for all three repeated measures ANOVA for investigating the simple effects of Environment in each of three music group, Tables 5.7–5.9 can be obtained.

Testing Sphericity Assumption Table 5.7 shows that the sphericity is not significant for the data in any of the three music group; hence, the sphericity assumption is not violated in any of the three music groups. Since the assumption is not violated, the significance of F for comparing the packaging performance of three environment groups in each music category shall be tested by using the p values mentioned in front of the "Sphericity Assumed" row in Table 5.8.

Testing Significance of F Values In this factorial design due to repeated measures single error variance could not be calculated for testing each simple effect; hence,

Table 5.7   Mauchly's Test of Sphericity in Different Music Categories

						Epsilon		
Music	Within-Subjects Effect	Mauchly's W	Approx. Chi-Square	df	Sig. (p value)	Greenhouse–Geisser	Huynh–Feldt	Lower-bound
No Music	Environment	0.483	2.914	2	0.233	0.659	0.802	0.500
Jazz	Environment	0.988	0.049	2	0.976	0.988	1.000	0.500
Instrumental	Environment	0.850	0.650	2	0.723	0.870	1.000	0.500

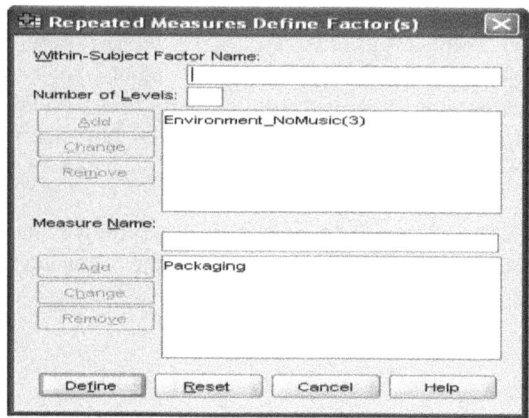

Figure 5.11   Screen showing options for defining independent and dependent variables

three separate one-way repeated measures ANOVA have been applied. Due to multiple ANOVA the familywise error rate 0.05 will be inflated approximately three times hence for testing the significance of F value for each repeated measures ANOVA the level of significance would be taken as 0.017(=0.05/3). A simple rule is to divide the familywise error rate (α) by the number of one-way repeated measures ANOVA applied.

Table 5.8 shows that the F values for the Environment in No music group ($p = 0.013$) and Instrumental group ($p = 0.003$) are significant because both these p values are less than 0.017. However, the F value for Jazz group ($p = 0.025$) is not significance because its p value is more than 0.017. Readers kindly note that the significance of these F values have been tested by comparing its associated p value with that of 0.017 but will be reported as significant at 5% level only.

Pairwise Comparisons of Means of Environment Groups in Each Music Group   Since F value is significant for the No music and Instrumental groups, the pairwise comparisons shall be investigated for these two groups only. From Table 5.9, it can be seen that there is a significant difference in the packaging performance of the subjects

ILLUSTRATION   113

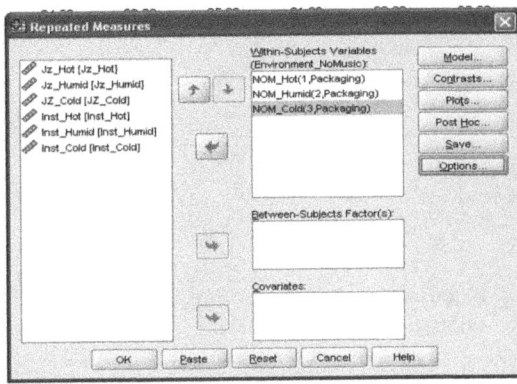

Figure 5.12 Screen showing option for selecting three levels of environment in no music group

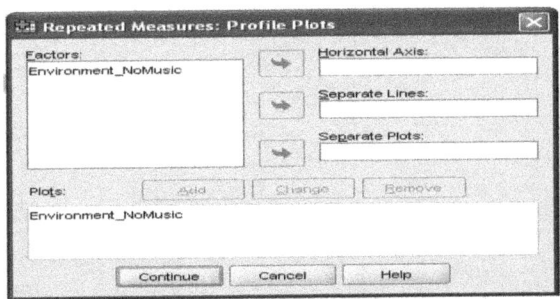

Figure 5.13 Screen showing option for means plot of Environment in no music group

Figure 5.14 Screen showing option for computing various outputs in one-way repeated measures ANOVA

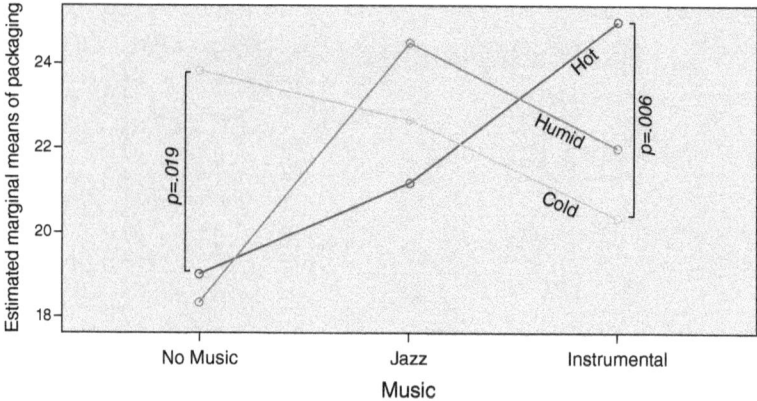

Figure 5.15 Marginal means plot of Music × Environment

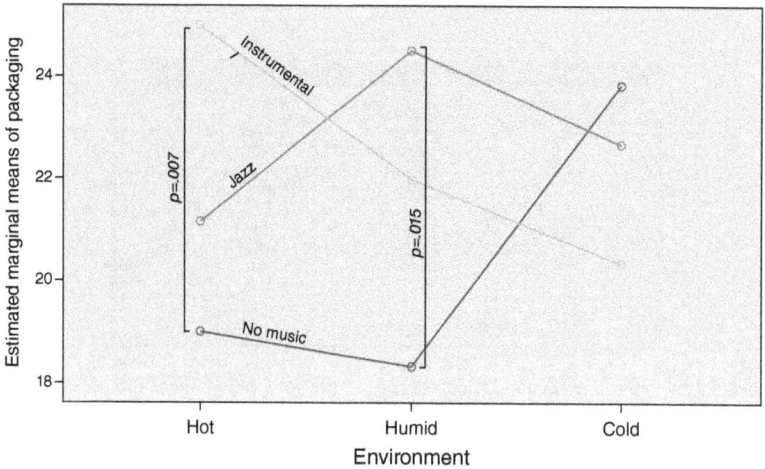

Figure 5.16 Marginal means plot of Environment × Music

while working in the hot and cold environments ($p < 0.017$) when there is no background music. Similarly with the instrumental music in the background also there is significant difference between the performance of the subjects in the hot and cold environment ($p < 0.017$). On the other hand, when Jazz music is in the background the performance of the subjects does not differ in all the three environments.

Marginal Means Plots (Music × Environment)   Results of Table 5.9 and the marginal means shown in Table 5.3 can be used to show the means plots of different environment groups in each music category as shown in Figure 5.15. This means plot

Table 5.8 F-table for Testing Significance of Environment (Within-Subjects) Effect in Each Music Category

Measure: Packaging

Source		Type III SS	df	Mean Square	F	Sig. (p value)	Partial Eta Squared
No Music							
Environment	**Sphericity Assumed**	108.111	2	54.056	7.000	**0.013**	0.583
	Greenhouse–Geisser	108.111	1.318	82.021	7.000	0.030	0.583
	Huynh–Feldt	108.111	1.605	67.369	7.000	0.021	0.583
	Lower-bound	108.111	1.000	108.111	7.000	0.046	0.583
Error (Environment)	Sphericity Assumed	77.222	10	7.722			
	Greenhouse–Geisser	77.222	6.590	11.717			
	Huynh–Feldt	77.222	8.024	9.624			
	Lower-bound	77.222	5.000	15.444			
Jazz							
Environment	Sphericity Assumed	33.444	2	16.722	5.473	0.025	0.523
	Greenhouse–Geisser	33.444	1.976	16.927	5.473	0.025	0.523
	Huynh–Feldt	33.444	2.000	16.722	5.473	0.025	0.523
	Lower-bound	33.444	1.000	33.444	5.473	0.066	0.523
Error (Environment)	Sphericity Assumed	30.556	10	3.056			
	Greenhouse–Geisser	30.556	9.879	3.093			
	Huynh–Feldt	30.556	10.000	3.056			
	Lower-bound	30.556	5.000	6.111			
Instrumental							
Environment	**Sphericity Assumed**	67.111	2	33.556	10.863	**0.003**	0.685
	Greenhouse–Geisser	67.111	1.739	38.585	10.863	0.005	0.685
	Huynh–Feldt	67.111	2.000	33.556	10.863	0.003	0.685
	Lower-bound	67.111	1.000	67.111	10.863	0.022	0.685
Error (Environment)	Sphericity Assumed	30.889	10	3.089			
	Greenhouse–Geisser	30.889	8.696	3.552			
	Huynh–Feldt	30.889	10.000	3.089			
	Lower-bound	30.889	5.000	6.178			

provides clear picture of the analysis. This plot indicates that the packaging performance of the subjects is significantly higher in the cold environment in comparison to that of hot and humid environments when there is no background music. However, this trend is reversed, when the back ground music is instrumental, the performance is better in hot climate in comparison to that of the cold climate.

Simple Effect of Music (Within-Subjects)   By using the procedure discussed above in generating the outputs for the simple effect of Environment the outputs for the three one-way repeated measures ANOVA for investigating the simple effects of Music can be obtained in SPSS. By combining the outputs for all the three repeated measures ANOVA, Tables 5.10–5.12 have been obtained.

Remark

1. For comparing the effect of Music in Hot environment, define factor name as Music_Hot in Figure 5.11 and select the following three variables in the screen shown in Figure 5.12.

   Hot_NoMusic

   Hot_Jz

   Hot_Inst

2. Similarly for comparing the effect of Music in Humid environment, define factor name as Music_Humid in Figure 5.11 and select the following three variables in the screen shown in Figure 5.12.

   Humid_NoMusic

   Humid_Jz

   Humid_Inst

3. Similarly for comparing the effect of Music in Cold environment, define factor name as Music_Cold in Figure 5.11 and select the following three variables in the screen shown in Figure 5.12.

   Cold_NoMusic

   Cold_Jz

   Cold_Inst

Testing Sphericity Assumption   Table 5.10 shows that the sphericity is not significant for the data of music groups in any of the three environment group; hence, the sphericity assumption is not violated in any environment group. The significance of F for comparing the packaging performance among the music groups in each environment group shall be tested by using the p value mentioned in front of the "Sphericity Assumed" row in Table 5.11.

Testing Significance of F Values   Table 5.11 shows that the F values for the Music in hot environment ($p = 0.000$) and in humid environment ($p = 0.011$) are significant because their associated p values are less than 0.017. The F value for the Music is not significant in cold climate.

ILLUSTRATION   117

Table 5.9   Pairwise Comparisons of Means in Each Music Category

					95% CI for Difference[a]	
(I) Environment	(J) Environment	Mean Difference (I–J)	Std. Error	Sig[a]	Lower Bound	Upper Bound
No Music						
Hot	Humid	0.667	1.498	1.000	−4.628	5.961
	Cold	−4.833[b]	1.078	**0.019**	−8.642	−1.025
Humid	Hot	−0.667	1.498	1.000	−5.961	4.628
	Cold	−5.500	2.078	0.137	−12.843	1.843
Cold	Hot	4.833[b]	1.078	**0.019**	1.025	8.642
	Humid	5.500	2.078	0.137	−1.843	12.843
Jazz						
Hot	Humid	−3.333	1.054	0.075	−7.059	0.392
	Cold	−1.500	0.957	0.534	−4.884	1.884
Humid	Hot	3.333	1.054	0.075	−0.392	7.059
	Cold	1.833	1.014	0.391	−1.750	5.416
Cold	Hot	1.500	0.957	0.534	−1.884	4.884
	Humid	−1.833	1.014	0.391	−5.416	1.750
Instrumental						
Hot	Humid	3.000	1.065	0.112	−0.762	6.762
	Cold	4.667[b]	0.803	**0.006**	1.830	7.504
Humid	Hot	−3.000	1.065	0.112	−6.762	0.762
	Cold	1.667	1.145	0.616	−2.380	5.713
Cold	Hot	−4.667[b]	.803	**0.006**	−7.504	-1.830
	Humid	−1.667	1.145	0.616	−5.713	2.380

Based on estimated marginal means
[a] Adjustment for multiple comparisons: Bonferroni.
[b] The mean difference is significant at the 0.017 level.

Table 5.10   Mauchly's Test of Sphericity[b] in Different Music

						Epsilon[a]		
Environment	Within Subjects Effect	Mauchly's W	Approx. Chi-Square	df	Sig. (p value)	Greenhouse– Geisser	Huynh– Feldt	Lower- bound
Hot	Music	0.585	2.148	2	0.342	0.706	0.903	0.500
Humid	Music	0.858	0.614	2	0.736	0.875	1.000	0.500
Cold	Music	0.439	3.296	2	0.192	0.640	0.764	0.500

Table 5.11 *F*-table for Testing Significance of Music (within-subjects) Effects in Each Environment

Measure: Packaging

Source		Type III SS	df	Mean Square	F	Sig. (p value)	Partial Eta Squared
Hot							
Music	**Sphericity Assumed**	110.778	2	55.389	22.557	**0.000**	0.819
	Greenhouse–Geisser	110.778	1.413	78.402	22.557	0.001	0.819
	Huynh–Feldt	110.778	1.806	61.345	22.557	0.000	0.819
	Lower-bound	110.778	1.000	110.778	22.557	0.005	0.819
Error(Music)	Sphericity Assumed	24.556	10	2.456			
	Greenhouse–Geisser	24.556	7.065	3.476			
	Huynh–Feldt	24.556	9.029	2.720			
	Lower-bound	24.556	5.000	4.911			
Humid							
Music	**Sphericity Assumed**	115.444	2	57.722	7.286	**0.011**	0.593
	Greenhouse–Geisser	115.444	1.751	65.936	7.286	0.016	0.593
	Huynh–Feldt	115.444	2.000	57.722	7.286	0.011	0.593
	Lower-bound	115.444	1.000	115.444	7.286	0.043	0.593
Error(Music)	Sphericity Assumed	79.222	10	7.922			
	Greenhouse–Geisser	79.222	8.754	9.049			
	Huynh–Feldt	79.222	10.000	7.922			
	Lower-bound	79.222	5.000	15.844			
Cold							
Music	Sphericity Assumed	38.111	2	19.056	2.074	0.176	0.293
	Greenhouse–Geisser	38.111	1.281	29.753	2.074	0.200	0.293
	Huynh–Feldt	38.111	1.529	24.930	2.074	0.192	0.293
	Lower-bound	38.111	1.000	38.111	2.074	0.209	0.293
Error(Music)	Sphericity Assumed	91.889	10	9.189			
	Greenhouse–Geisser	91.889	6.405	14.347			
	Huynh–Feldt	91.889	7.644	12.021			
	Lower-bound	91.889	5.000	18.378			

ILLUSTRATION   119

Table 5.12   Pairwise Comparisons of Means in Each Environment

(I) Music	(J) Music	Mean Difference (I-J)	Std. Error	Sig[a]	95% CI for Difference[a] Lower Bound	Upper Bound
Measure: Packaging						
Hot						
No music	Jazz	−2.167	0.543	0.031	−4.084	−0.249
	Instrumental	−6.000[b]	1.065	**0.007**	−9.762	−2.238
Jazz	No music	2.167	0.543	0.031	0.249	4.084
	Instrumental	−3.833	1.014	0.039	−7.416	−0.250
Instrumental	No music	6.000[b]	1.065	**0.007**	2.238	9.762
	Jazz	3.833	1.014	0.039	0.250	7.416
Humid						
No music	Jazz	−6.167	1.302	**0.015**	−10.767	−1.566
	Instrumental	−3.667	1.687	0.245	−9.627	2.294
Jazz	No music	6.167	1.302	**0.015**	1.566	10.767
	Instrumental	2.500	1.839	0.697	−4.001	9.001
Instrumental	No music	3.667	1.687	0.245	−2.294	9.627
	Jazz	−2.500	1.839	0.697	−9.001	4.001

Based on estimated marginal means
[a] Adjustment for multiple comparisons: Bonferroni.
[b] The mean difference is significant at the 0.017 level.

Table 5.13   Test of Normality

Treatment groups	Shapiro–Wilk statistic	df	Sig.
NOM_Hot	0.960	6	0.820
NOM_Humid	0.907	6	0.415
NOM_Cold	0.830	6	0.108
Jz_Hot	0.928	6	0.566
Jz_Humid	0.803	6	0.062
JZ_Cold	0.918	6	0.493
Inst_Hot	0.853	6	0.167
Inst_Humid	0.879	6	0.266
Inst_Cold	0.905	6	0.404

Pairwise Comparison of Means of Music Groups in Each Environment Group   Since F value for Music is significant in hot as well as in humid environments, the pairwise comparisons of means shall be investigated in these two groups only. Table 5.12 reveals that there is a significant difference in the packaging performance of the subjects in the no music and instrumental music groups ($p < 0.017$) while working in the hot climate. Similarly packaging performance differs significantly in the no music and Jazz groups ($p < 0.017$) while working in humid climate.

Marginal Means Plots (Environment × Music) By combining the results of the post hoc test shown in Table 5.12 and the marginal means shown in Table 5.3 the marginal means plots of different music groups in each environment category can be obtained as shown in Figure 5.16. However, this output is generated by the SPSS during main analysis. This means plot gives clear picture of the analysis.

This plot indicates that the packaging performance of the subjects is significantly higher with the instrumental music in the background in comparison to that of when there is no music while working in the hot environment. However, this trend is not observed when the subjects work in humid and cold environment. While working in the humid environment subject's packaging performance is significantly higher with the Jazz music in the background in comparison to that when there is no music. However, no significant difference in the performance has been observed in all the three music groups when the subjects work in the cold environment.

How to Report the Findings

In this illustration, two-way repeated measures design was used to investigate the main effects of Music, Environment, and the interaction between them. The hypotheses were tested at the significance level of 0.05.

Assumptions In order to test the suitability of the design its required assumptions were tested which are reported as follows:

- Since both the independent variables (music and *environment)* were categorical and dependent variable *(packaging performance)* was a metric variable, the assumption about the data type was not violated.
- Since subject's performance was tested independently with each other, the assumptions of independence also was not violated.
- Since the Shapiro–Wilk statistic was not significant in each of the nine group of scores, the assumption of normality was not violated (Table 5.13).
- Mauchly's W test was significant in Music and the estimated value of Epsilon (ε) was less than 0.75; hence, the Greenhouse–Geisser correction was applied to the degrees of freedom while testing the significance of F for music. However, Mauchly's W test was nonsignificant for Environment and Interaction; hence, no correction was applied for testing the significance of these effects (Table 5.4).

Testing Main Effects

- Since F value for Music (Greenhouse–Geisser) and Environment (Sphericity Assumed) was nonsignificant, the null hypothesis that the average packaging performance is same in all the three music groups irrespective of environment categories was not rejected at 5% level. Similarly, the performance of the subjects did not differ in all the three environment groups irrespective of the music background (Table 5.4).

ILLUSTRATION   121

Testing Simple Effects

- Since *F* value for the interaction (Environment × Music) was significant, the simple effects of music as well as environment were further investigated.
- Since sphericity assumption was not violated for the data in each of the three music group for comparing performance among different environment groups, no correction was applied (Table 5.7). The *F* value for Environment in No music ($p = 0.013$) as well as in Instrumental ($p = 0.003$) groups was significant; hence, the null hypothesis of no difference in mean packaging scores among the three environment groups in No music group as well as in instrumental group was rejected at 5% level (Table 5.8). The pairwise comparisons of means suggested the following:
- The subject's performance improved in cold environment in comparison to that of the hot and humid environments when there was no music in the background.
- When instrumental music was there in the background, the subject's performance increased in hot environment significantly in comparison to that of the cold environment.
- When Jazz music was used in the background, no difference was observed in the performance of subjects in different environmental groups.
- Since sphericity assumption was not violated for the data in each of the three environment groups for comparing performance among different music groups, no correction was applied (Table 5.10). The value of *F* for Music was significant in Hot climate($p = 0.000$) as well as in Humid climate($p = 0.011$); hence, the null hypothesis of no difference in mean packaging scores among the three music groups was rejected at 5% level in hot as well as in humid environment category (Table 5.11). The pairwise comparison of means suggested the following:
- In hot environment, the performance of the workers increased if instrumental music was provided in the background in comparison to when there was no music or Jazz music.
- In humid environment, performance of the workers was significantly higher when Jazz music was there in the background instead of no music.
- In cold environment, no difference was observed in all the three different types of music intervention.

Inference

On the basis of the sample observations, it may be concluded that if no background music was provided to the workers their performance increased in the cold environment in comparison to hot and humid environments. But this trend was reversed if the background music was instrumental where hot climate was more productive. Further, while working in hot environment, instrumental music enhanced the performance of the workers in comparison to that when no music was allowed. However, while working in the humid environment performance of the workers was enhanced with the Jazz music in the background.

EXERCISE

5.1. Explain all assumptions used in two-way repeated measures design. What happens if these assumptions are violated? As a researcher what would you do if normality assumption is violated and there is no way to rectify it in your study?

5.2. A two-way repeated measures design was organized to investigate the effect of coffee drinking on the power of concentration over a period of time. Five subjects were tested for their concentration power before the experiment, after one and two weeks of the coffee consumption. After considerable period of time the same subjects were tested at these time points while being under placebo treatment. Describe layout design in the study.

5.3. While solving two-way repeated measures design with SPSS what steps are followed? What are the outputs that need to be generated?

5.4. In a two-way repeated measures design if factors A and B have three and four levels, respectively, whereas six subjects in the sample have been tested under all A × B treatment conditions, how the total sum of squares shall be distributed into different components along with degrees of freedom. Show it by a schematic diagram.

5.5. What do you mean by main and simple effects? Explain by means of an example. Which effect is more important to investigate and why in a factorial experiment.

ASSIGNMENT

5.1. A nutritionist wanted to investigate the effect of green tea and time duration on the weight loss of housewives. He organized a two-way repeated measures design in which 5 housewives participated. The three treatments T1(green tea with 1800 calorie diet), T2(green tea with 2000 calorie diet), and T3(placebo) were given to these subjects in a randomized fashion. Each subject received all three treatments for 4 weeks and was tested for their weight before starting treatment and after two and four weeks during treatment. Sufficient time gap was given between any two treatments so as to done away the effect of previous treatment and bring back nearly the same weight of the participants. The data so obtained are shown in the Table 5.14. Apply two-way repeated measures design and do the following:
   a. Show different components of the total sum of squares along with their respective degrees of freedom in this repeated measures design.
   b. Construct various hypotheses which need to be tested.
   c. Discuss your findings by taking family wise error rate as 0.05.

5.2. A study was organized by a psychologist to see the effect of beverages and time of testing on reaction time. Five male subjects were randomly selected

Table 5.14   Data on Weight (in kg) Obtained on the Housewives Under Different Treatment Conditions

	Time		
	Zeroth week	Two weeks	Four weeks
Green tea + 1800C (T1)	60	56	54
	65	60	59
	62	57	56
	67	62	60
	59	56	55
Green tea + 2000 (T2)	60	58	54
	65	64	61
	62	60	57
	67	66	64
	60	59	55
Placebo (T3)	61	60	61
	65	64	64
	62	62	61
	67	66	67
	59	59	59

Table 5.15   Data on reaction time (in msec) obtained on subjects under different treatment conditions

Time of Testing		
Morning	Afternoon	Evening
Coffee		
250	225	210
280	250	260
290	240	250
270	245	250
280	265	250
Tea		
200	235	215
180	225	195
170	210	200
180	225	195
190	210	200
Placebo		
215	225	200
200	220	190
210	220	205
220	235	205
205	215	195

for the study. Each subject was tested for his reaction time after consuming coffee, tea, and placebo drinks in the morning, afternoon and evening sessions. Counterbalancing was done to nullify the order effect. The data so obtained are shown in Table 5.15. Apply two-way repeated measures design to test the main and simple effect. Discuss your findings by drawing the means plot.

BIBLIOGRAPHY

Anscombe FJ. The validity of comparative experiments. J Roy Stat Soc A 1948;111(3):181–211. DOI: 10.2307/2984159JSTOR 2984159. MR 30181.

Bailey RA. *Design of Comparative Experiments*. Cambridge University Press; 2008. ISBN: ISBN 978-0-521-68357-9. Pre-publication chapters are available on-line.

Bakeman . Recommended effect size statistics for repeated measures designs. Behav Res Methods 2005;37(3):379–384. DOI: 10.3758/bf03192707.

van Belle G. *Statistical Rules of Thumb*. 2nd ed. Hoboken, NJ: Wiley; 2008. ISBN: ISBN 978-0-470-14448-0.

Caliński T, Kageyama S. *Block Designs: A Randomization Approach, Volume I: Analysis. Lecture Notes in Statistics 150*. New York: Springer-Verlag; 2000. ISBN: ISBN 0-387-98578-6.

Cox DR, Reid NM. *The Theory of Design of Experiments*. Chapman & Hall/CRC; 2000. ISBN: ISBN 978-1-58488-195-7.

Kreuger C, Tian L. A comparison of the general linear mixed model and repeated measures ANOVA using a dataset with multiple missing data points. Biol Res Nurs 2004;6:151–157.

Lentner M, Bishop T. *Experimental Design and Analysis*. 2nd ed. P.O. Box 884, Blacksburg, VA 24063: Valley Book Company; 1993. ISBN: 0-9616255-2-X.

Muller KE, Barton CN. Approximate power for repeated -measures ANOVA lacking sphericity. J Am Stat Assoc 1989;84(406):549–555. DOI: 10.1080/01621459.1989.10478802.

Olejnik S, Algina J. Generalized eta and omega squared statistics: measures of effect size for some common research designs. Psychol Methods 2003;8:434–447. DOI: 10.1037/1082-989x.8.4.434.

6

TWO-WAY MIXED DESIGN

INTRODUCTION

Two-way mixed design is used to analyze the effect of two independent factors on some dependent variable in a situation where one of the factors is a between-subjects and the other is a within-subjects. Different treatments of within-subject factor are randomly assigned to all the subjects in each level of the between-subjects factor. In this design treatments are assigned by using randomization and repeated measures concept. Two-way mixed design is also known as split plot design. The purpose of using this design is to test the differences between two or more independent groups, while subjects are repeatedly measured on some dependent variable in each level of the within-subjects factor.

In two-way mixed design subjects in each level of the between-subjects factor are tested repeatedly under each treatment conditions of the within-subjects factor. Consider an experiment in which it is desired to compare the effect of different time of the day on the memory retention among boys and girls. Here, gender is a between-subjects factor having two levels (male and female) and the time is a within-subjects factor having three levels (morning, afternoon, and evening). In this experiment subjects would be tested for their memory retention in all the three treatment conditions (morning, afternoon, and evening). Here researcher is basically interested in three types of hypotheses; firstly, whether memory retention performance differs in three time periods irrespective of the gender; secondly, whether memory retention performance differs among boys and girls irrespective of the time of testing; and thirdly, whether interaction effect between gender and time of testing is significant.

Repeated Measures Design for Empirical Researchers, First Edition. J. P. Verma.
© 2016 John Wiley & Sons, Inc. Published 2016 by John Wiley & Sons, Inc.

If interaction exists it simply means that the pattern of memory retention during different time periods of the day differs in male and female. The effect of time and that of gender are known as main effects, whereas effect of time in each gender and that of effect of gender in each time condition are known as simple effects. The simple effects are investigated only when the interaction is significant. The main effects are meaningful only when the interaction is not significant. In this design order effect is an issue which is tackled by counterbalancing. Counterbalancing is done by dividing the subjects and exposing them in different treatment conditions randomly. Readers will understand it better by looking to the design format of the illustration discussed in this chapter.

In two-factor mixed design it is not necessary that different levels of the within-subjects factor are different treatments; it may be different durations as well. For instance, to compare the effect of three different types of diet plan on the bone density, an experiment may be carried out in which three groups of subjects may be kept on different diet plans for four months. In each diet group subjects may be tested initially, after two and four months. Here the levels of within-subjects factor are different time durations instead of different treatments. In this case there is no issue of order effect, hence no point of counterbalancing, because the effect in each diet group is cumulative. But then this is what the objective of the experiment is where one wishes to investigate the pattern of increase in the bone density in different durations in each of the three diet groups.

A particular case of the two-way mixed design is a pre-post-control design. This is a powerful design which is commonly used by researchers. In this design subjects in the sample is divided into two groups. The two groups are formed by randomly allocating treatments to the subjects. One group serves as experimental, whereas the other acts as control. The researcher is basically interested in the experimental group, whereas the control group is taken as base line. In experimental group all subjects are exposed to the treatment and are given a pre-test and post-test. On the other hand, in control group the subjects are pre-tested and post-tested by giving placebo treatment. The placebo is a fake treatment which has nothing to do with the experiment but is given to the subjects in the control group in order to make them feel that they are also a part of experiment. Another purpose of giving placebo treatment to the subjects is to ensure that they do not come to know whether they are under experimental or control group. This is done to reduce the bias and maintain seriousness in the experiment by the participants. Consider an experiment in which we are interested in examining the effect of a new herb solution on the hemoglobin contents. We may identify similar level of anemic subjects and randomly divide them into two groups. First group of subjects may be given the new herb for a certain period of time say twelve weeks, whereas the second group may serve as a control and be given a placebo for the same period of experimentation. In this case the treatment (herb and placebo) is a between-subjects factor and the time (pre and post) is a within-subjects factor. Since this pre-post control design is a specific case of the mixed design, it may be solved by using the two-factor mixed ANOVA, but a better way to solve this design would be to use the analysis of covariance.

ADVANTAGE OF TWO-WAY MIXED DESIGN

One of the main advantages of the two-way mixed design is that it allows investigating interaction effect between within-subjects and between-subjects factors. In this design between-subjects factor can be considered as a covariate. This design is efficient in comparison to that of single factor repeated measures design because part of the within-subjects factor variability is explained by the between-subjects factor, thereby reducing the error substantially. Due to reduction of within-subjects error, the two-way mixed design is very sensitive in detecting even the slightest variation in the levels of within-subjects factor. In repeated measures design post-hoc test for comparison of mean is not possible, but in mixed design due to the presence of between-subjects factor post-hoc test can be applied for pair-wise comparison of means among the levels of between-subjects factor.

ASSUMPTIONS

In order to use this design following assumptions should be tested. If one or more of these assumptions are not satisfied, the type I error inflates in testing various hypotheses in the study.

a. *Assumption on the data type*. The two independent variables (between-subjects and within-subjects) must be categorical and should have at least two categories, whereas the dependent variable should be measured on either interval or ratio scale.

b. *Assumption of normality*. Distribution of the data obtained on dependent variable must be normally distributed in each cell. For example in the 2×3 mixed design the data must be normally distributed in all six cells.

c. *Independence of observations*. Participant's response must be independent to each other.

d. *Homogeneity of variance*. The error variance of the dependent variable is equal across the levels of between-subjects factor in each level of the within-subjects factor. For instance, in 2×3 mixed design the variances of the data in both the levels of the between-subjects factor should be same in each of the three levels of the within-subjects factor. This can be tested by using the Levene's test in SPSS.

e. *Sample size*. The sample size in each cell should be preferably 20 or more. This enhances the power in the design and protects the violation of the two assumptions; normality and homogeneity of variances.

f. Sphericity. The variances of the differences between all combinations of within-subjects factor must be equal. In other words, correlations among the repeated measures are all equal. The sphericity assumption is tested by Mauchly's test and if this test is significant the sphericity assumption is said to be violated. In that case Greenhouse–Geisser or Huynh–Feldt correction is normally used depending upon the severity of the sphericity violation.

g. *Homogeneity of variance–covariance matrices.* The pattern of inter-correlations among the various levels of the within-subjects factor should be similar in different levels of the between-subjects factor. In other words, variance–covariance matrices of the dependent variable should be equal across the cells in all levels of the between-subjects factor. This assumption is tested by using the Box's M test in SPSS. This statistic is very sensitive, hence it is suggested that the level of significance should be used as 0.001 for testing the significance of this test. Thus, for homogeneity of variance–covariance matrices the Box's M test should be non-significant. In other words, homogeneity exists if p value is greater than 0.001. By testing this assumption, one ensures that the vector of the dependent variable follows a multivariate normal distribution.

APPLICATION

Two-way mixed design has many applications in different disciplines. Some of the applications are listed below:

1. A human resource manager may investigate the effect of three different intervention of training (onsite, offsite, and mix of these two) on learning skills for their employees. It has been observed that the women employees generally don't prefer outstation training, hence to investigate this fact also the investigator may take gender as a between factor. Here intention of the investigator is to compare the learning skills in three different modes of training as well as between male and female subjects. Simultaneously investigator may test the significance of interaction effect between training modes and gender as well. If researcher's interest is only to compare the effectiveness of three different modes of training, then in that case gender may be considered as a covariate and treated as a blocking variable.

2. A psychologist may like to investigate the effect of three different cognitive therapies on the stress level. He can do so by taking gender as between-subjects factor and levels of the cognitive therapy as a treatment factor (within-subjects factor). In this case the researcher's interest is to compare the effect of different cognitive therapies only, whereas gender variable has been introduced to reduce the experimental error. Since the analysis also provides the value of F for the gender, the stress levels of the male and female can be compared across the treatment levels as well.

3. An educational psychologist may like to investigate the effect of learning methods (traditional, audio-visual, and self learning) and IQ (high and low) on memory retention. In traditional learning group subjects may be asked to recall as many objects as they can and write them on the blackboard within a fixed time. In audio-visual group objects may be shown to the subjects through audio-visual aid and then may be asked to recall. However, in self-learning method the subjects may be given the number of objects to see them physically and then recall. To address the research issues the educational psychologist may plan a two-way mixed design by randomly selecting the subjects in the

high and low IQ groups. Here IQ is a between-subjects factor, whereas learning method is a within-subjects factor because all the subjects in each of the two IQ groups are repeatedly tested in all the three treatment conditions (traditional, audio-visual, and self learning).

4. A basketball coach may wish to investigate the effect of shooting distance (3, 4 and 5 m) and gender on shooting performance in basketball. Here distance is a within-subjects and gender is a between-subjects factor, respectively.

5. A nutritionist may be interested to compare the effect of three diet programmes on weight reduction in a six-week experiment. Since nutritional requirement depends upon the activity level of the subjects, activity level may be considered as a covariate. The subjects may be classified in different blocks of active, semi-active, and sedentary subjects and then in these blocks the treatments (three different types of diets) may be randomly allocated. Further, it may be of interest to the researcher to find the optimum duration in which the significant reduction of weight can be achieved. To fulfill these objectives a two-way mixed design can be planned. In each diet group subjects may be tested for their weight initially, after two and four weeks during observing the diet plans. Here diet is a between-subjects factor, whereas the time is a within-subjects factor.

LAYOUT DESIGN

In mixed design one of the factors is between-subjects, whereas the other is a within-subjects. Layout of the mixed design depends upon whether levels of the within-subjects factor are different treatment conditions or time periods. The layout of the mixed design in these two situations has been discussed in the following sections.

Case I: When Levels of the Within-Subjects Factor are Different Treatments

In this mixed design all subjects in each level of between-subjects factor are tested under each treatment of within-subjects factor. One must ensure that there is no learning effect due to testing in the earlier treatment(s). To ensure this there should be sufficient time gap between implementing any two treatments. The other issue in such studies is that of the order effect. Testing all subjects under different treatments in a specific sequence may affect the subject's performance due to learning or fatigue. This effect is known as order effect. The order effect is tackled by using counterbalancing in the design. This is done by dividing the sample into c groups in each level of between-subjects factor, where c is the number of levels of the within-subjects factor. Let us suppose that within-subjects factor has three levels, then the first group will undergo the first treatment condition thereafter second and then the third, whereas the second group may undergo the second treatment condition first and then the first and later the third in the sequence. Similarly

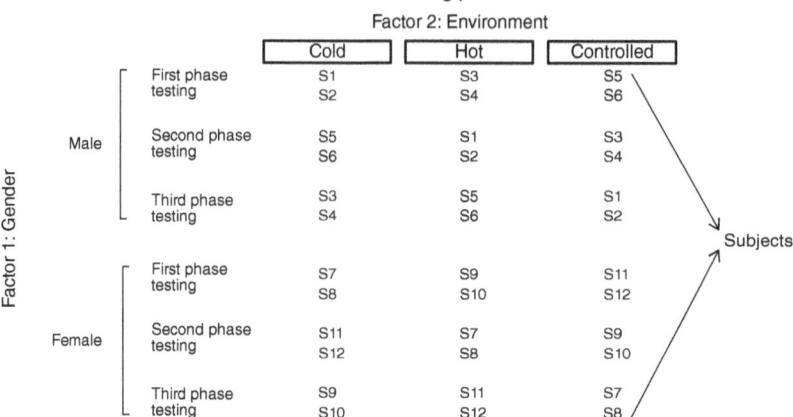

Figure 6.1 Layout of the mixed design with two-factors where levels of the within-subjects factor are the three treatments

the order of the third group may be the third treatment condition first then the second and thereafter the first. The same protocol is followed by the subjects in each level of the between-subjects factor.

Consider an experiment in which it is desired to investigate the effect of three different environmental conditions (cold, hot, and controlled) on mood behavior of the subjects (male and female). Here gender is between-subjects factor, whereas environment is within-subjects. If sample consists of six male and six female subjects, layout of the mixed design can be shown by the Figure 6.1. As per the protocol in the first phase of testing the subjects S1 and S2 will undergo mood test under cold climate, the subjects S3 and S4 will be tested under hot climate, and the subjects S5 and S6 will get tested under the controlled climate. Similarly the testing protocol for the subjects during second and third phase of testing has been shown in the Figure 6.1.

Case II: When Levels of the Within-Subjects Factor are Different Time Durations

In this layout all subjects in each level of the between-subjects factor are repeatedly tested at different time points during treatment. Here performance of subjects in particular time duration is affected due to the cumulative learning effect in the earlier durations, but that's what the objective of the study is. In such studies a researcher is interested to know the behavior of subjects due to treatment over a period of time. Consider an exercise therapy programme in which the participants (male and female both) are tested for their weight before experiment, after 2 weeks, 4 weeks, and 6 weeks during treatment intervention. If the sample consists of five male and five female subjects, the layout of this mixed design can be shown by the Figure 6.2.

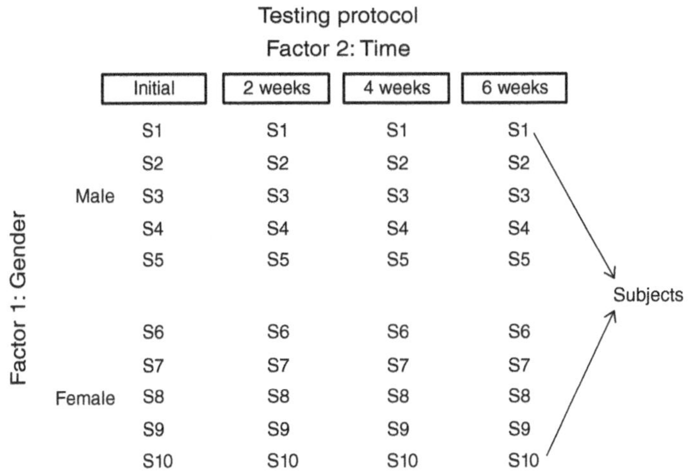

Testing protocol

Factor 2: Time

Initial	2 weeks	4 weeks	6 weeks

Factor 1: Gender

	Male	S1	S1	S1	S1
		S2	S2	S2	S2
		S3	S3	S3	S3
		S4	S4	S4	S4
		S5	S5	S5	S5
		S6	S6	S6	S6
		S7	S7	S7	S7
	Female	S8	S8	S8	S8
		S9	S9	S9	S9
		S10	S10	S10	S10

Subjects

Figure 6.2 Layout of the mixed design with two factors where levels of the within-subjects factor are the time durations

STEPS IN SOLVING MIXED DESIGN WITH TWO-WAY ANOVA

In mixed design a researcher is interested to investigate the effect of within-subjects and between-subjects factors on some dependent variable. Simultaneously an inter-action effect between both the factors is also investigated. Here comparison among the levels of within-subjects factor is done by using the concept of repeated measures design, whereas the comparison among the levels of the between-subjects factor is done by using the independent measures design. This is so because the same subjects are tested in all the levels of the within-subjects factor, whereas different subjects are tested in each level of the between-subjects factor. Following steps are used in solving a mixed factor design:

1. Test the assumption of normality for the scores obtained on dependent variable in different treatments for each level of the between-subjects factor as discussed in Chapter 3.
2. Describe the layout of design
3. Write research questions which need to be answered
4. Construct hypotheses to be tested
5. Decide family-wise error rate (α) at which the hypotheses need to be tested
6. Use SPSS commands to generate the following outputs
   a. Descriptive statistics including mean and standard deviation
   b. Box's test for homogeneity of variance–covariance matrices
   c. Mauchly's test of Sphericity including estimated value of epsilon(ε)
   d. Table of F values for testing within-subjects effects and interaction effect
   e. Levene's test of equality of variances
   f. Table of F for testing between-subjects effects

g. If F for within-subjects effect is significant, select pair-wise comparison table using *Bonferroni* correction.

h. If F for between-subjects effect is significant, select pair-wise comparison table using Tukey HSD post-hoc test.

i. Means plot for each factor (A and B)

j. Means plot for interaction ($A \times B$)

k. Means plot for interaction ($B \times A$)

7. Test the assumption of homogeneity of variance–covariance matrices by Box's M test.

8. Test sphericity assumption

9. If sphericity is significant apply correction (Greenhouse–Geisser or Huynh–Feldt) and test the significance of F value for within-subjects factor and interaction at the corrected degrees of freedom, otherwise use the p-value for testing the significance of F by assuming sphericity.

10. If F value for the within-subjects effect is significant, apply paired t-test for pair-wise comparisons among group means by applying Bonferroni correction.

11. If F-ratio for the between-subjects effect is significant, apply Tukey post-hoc test for comparing group means.

12. If Interaction effect is significant, test simple effects of between-subjects as well as within-subjects factor.

13. Report the findings.

The procedure involved in solving the mixed design shall be explained by means of the below-mentioned illustration.

ILLUSTRATION

A market analyst wanted to investigate the effect of age and movie types on the audience enjoyment. A research study was planned in which eighteen subjects were randomly selected, six each from the teens, mid age, and old age categories. All these eighteen subjects were shown all the three selected blockbuster movies in the experiment by using counterbalancing technique. Out of the three movies selected the first was based on romantic, the second on social theme, and the third was an action thriller. After viewing each movie the subjects were asked to rate the enjoyment on a ten point scale. The value 1 indicated the worst movie ever seen, creating headache and not recommending to anyone, whereas the score 10 indicated the best movie for enjoyment which provided an ultimate pleasure to the audience. Enjoyment scores of the subjects so obtained are shown in the Table 6.1

Layout Design

In this study effect of age and movie is to be seen on enjoyment. Since age is a between-subject factor and movie is a within-subject, the study may be planned using mixed design layout which can be shown by the Figure 6.3.

ILLUSTRATION 133

Table 6.1 Score on Enjoyment Reported by the Subjects after Watching Movies

		Subject	Movie		
			Romantic	Social	Action
Age	Teens	1	65	50	57
		2	65	56	62
		3	59	46	53
		4	67	50	54
		5	66	52	60
		6	62	51	63
	Mid age	7	65	62	48
		8	60	67	53
		9	57	52	44
		10	61	55	43
		11	62	64	46
		12	62	65	47
	Old age	13	61	67	50
		14	58	62	52
		15	62	68	46
		16	60	66	48
		17	55	65	53
		18	60	72	56

Testing protocol
Factor 2: Movie

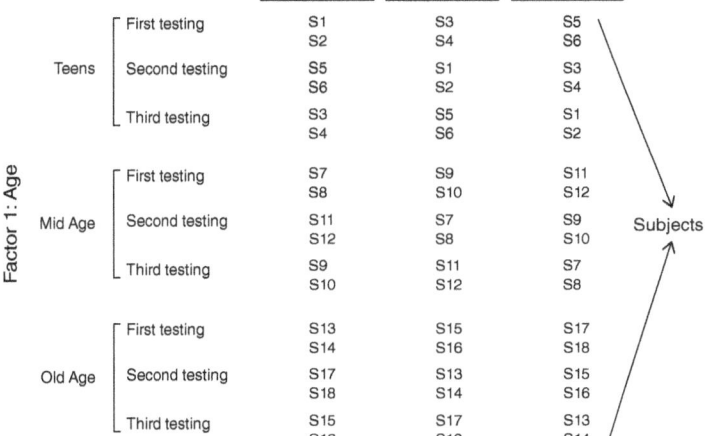

Figure 6.3 Layout of the mixed design in the illustration

If you look into the design in Figure 6.3, all subjects in each age category undergo all the treatment conditions but not in sequence. For instance, the subjects S1 and S2 are first exposed to the movie on *romantic*, then *social,* and thereafter the *action*. On the other hand, the subjects S3 and S4 are first exposed to the movie on social theme, then on action, and thereafter on romantic. This counterbalancing has been done in order to avoid the order effect on the subject's response.

Distribution of Variation and Degrees of Freedom

Let's see how the total variation in this mixed design is distributed in different components. If you look into the data of the Table 6.1, the total variability (SS) is because of the variation between subjects ($SS_{Subjects}$) and variation within subjects ($SS_{Within\_subjects}$). Further variation between subjects is because of the variation due to age factor (SS_{Age}) and variation between subjects within each age category (SS_{Error_Age}). On the other hand, variation within subjects is because of three components, that is, variation due to Movie factor (SS_{Movie}), variation due to interaction between Age and Movie ($SS_{Age \times Movie}$), and variation due to interaction of movie $\times$ subjects within-subjects in all the levels of factor Age (SS_{Error_Movie}).

In this illustration Age is a between-subjects factor and Movie is a within-subjects and therefore the distribution of total variation in this mixed design shall be as follows:

$$\text{Total } SS = SS_{Subjects} + SS_{Withing \text{ } Subjects}$$

$$= (SS_{Age} + SS_{Error\_Age}) + (SS_{Movie} + SS_{Age \times Movie} + SS_{Error\_Movie})$$

where

r	=	number of levels of Movie factor (within-subjects) = 3
c	=	number of levels of Age factor (between-subjects) = 3
n	=	number of subjects in each of the r levels of Age factor = 6

and

SS_{Age}	=	sum of squares between 3 levels of Age factor
SS_{Error_Age}	=	Error Between = pooled sum of squares between subjects within each level of the factor Age
SS_{Movie}	=	sum of squares between 3 levels of factor Movie
$SS_{Age \times Movie}$	=	sum of squares due to the interaction of factors Age and Movie
SS_{Error_Movie}	=	Error Within = pooled sum of squares due to interaction of movie $\times$ subjects within-subjects in all the levels of factor Age

ILLUSTRATION   135

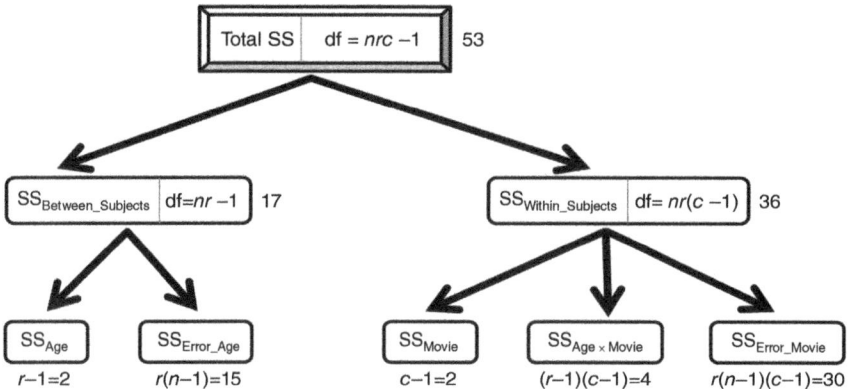

Figure 6.4   Scheme of distributing total sum of squares and degrees of freedom in the mixed design

In this illustration of mixed design the total degrees of freedom, that is $nrc-1(=$ 53), have been partitioned into $nr-1(=17)$ df due to variation between subjects and $nr(c-1)(=36)$ df due to variation within subjects. Further, $nr-1(=17)$df due to variation between subjects is partitioned into $r-1(=2)$ df for Age and $r(n-1)(=15)$ df for Error_Age (variation among subjects within each age category). On the other hand, $nr(c-1)(=36)$ df of within subjects variation has been partitioned into $c-1(=2)$ df for Movie, $(r-1)(c-1)(=4)$ df for Age × Movie, and $r(n-1)(c-1)(=30)$ df for Error_Movie (variation due to interaction of movie × subjects within-subjects in all the levels of factor Age). The distribution of Total SS into different components along with their degrees of freedom has been shown graphically in the Figure 6.4.

Research Questions

In this illustration the researcher is basically interested to investigate the below-mentioned three research questions. Out of these the research questions 'a' and 'b' can be investigated through the main effects, whereas 'c' may be investigated by analyzing simple effects.

a. Whether enjoyment in watching movie depends upon the age of the subjects.
b. Whether enjoyment in watching movie depends upon the type of movie seen by the subject.
c. Whether interaction between age and movie type affects the enjoyment in watching movie. In other words, whether the pattern of response on enjoyment in watching different types of movies differs in different age categories. And similarly whether the response pattern on enjoyment in different age categories differs in each movie category.

Hypothesis Construction

To investigate the above-mentioned research questions following hypotheses shall be tested:

a. *Main effect of Movie*

$$H_0 : \mu_{Romantic} = \mu_{Social} = \mu_{Action}$$

against $\quad$ H_1 : At least one group mean differs

b. *Main effect of Age*

$$H_0 : \mu_{Teens} = \mu_{Mid\_age} = \mu_{Old\_age}$$

against $\quad$ H_1 : At least one group mean differs

c. *Interaction effect (Age × Movie)*

H_0 : There is no interaction between Age and Movie

against $\quad$ H_1 : The interaction between Age and Movie is significant

Testing first two hypotheses shall investigate the first two research questions (main effects) mentioned above, whereas the third research question can be answered by testing the hypothesis on the interaction effect. Kindly note that testing of two main effects shall be meaningful only if the interaction effect is non-significant. In case the interaction effect is significant, testing these main effects becomes meaningless. In that case, simple effects are investigated. Simple effect refers to comparing enjoyment scores of three age categories in each level of the movie and comparing subject's response on three different types of movies in each age category.

Level of Significance

The family-wise error rate (α) in this illustration shall be taken as 0.05. This is a mixed design; hence if the value of F for between-subjects factor is significant, the pair-wise comparison of group means among different levels of between-subjects factor shall be done by using the post-hoc test as is done in case of independent measures design. Since post-hoc test for the repeated measures does not exist, if the F value for the within-subjects factor (Movie) is significant, a Bonferroni correction shall be applied for correcting the level of significance. In the mixed design no single error sum of squares is computed unlike independent measures design; hence if the interaction effect is significant, simple effects of both the factors (within-subjects as well as between-subjects) shall be evaluated by applying three one-way repeated measures ANOVA for within-subjects factor and three independent measures ANOVA for between-subjects factor. Thus, the significance of F for Age in each type of movie shall be tested at 0.017 (= 0.05/3) level. Similarly in testing the effect of Movie in each age category, significance of F will be tested at the significance level 0.017.

ILLUSTRATION   137

Solving Mixed Design with Two-Way ANOVA using SPSS

Solution of this Mixed Design with two-way ANOVA shall be discussed by using the SPSS software. To use this software a data file needs to be prepared. Since the procedure for preparing data file has already been explained in Chapter 3, it will not be discussed here. First time users of SPSS are advised to go through the Chapter 3 before following the procedure discussed here. After preparing data file, follow the sequence of commands mentioned below as shown in the Figure 6.5.

After clicking on **Repeated Measures Design** command, the screen shown in Figure 6.6 shall be obtained for defining independent and dependent variables. By default the "Within-Subject Factor Name" is written as factor 1. Replace this by *Movie*, the independent variable in this illustration. Type the number of levels as 3 as there are three different levels of the movie on which the data has been obtained. Click on **Add**. Type the name of dependent variable in the "Measure Name" area as *Enjoyment*. Click on **Add**. You will get the screens as shown in Figures 6.6 and 6.7 before and after clicking on the **Add** command, respectively.

Click on the command **Define** in the screen shown in Figure 6.7 for getting the screen for getting the option for selecting levels of the Within-subjects factor as shown in Figure 6.8. Select three within-subjects variables (romantic movie, social movie,

Analyze ⟶ General minear model⟶ Repeated measures

Figure 6.5   Screen for initiating commands for the mixed design

Figure 6.6   Screen showing options for defining dependent and independent variables and its levels

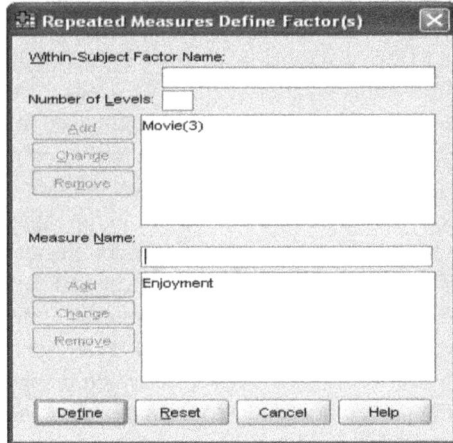

Figure 6.7 Screen showing options for adding independent and dependent variables for analysis

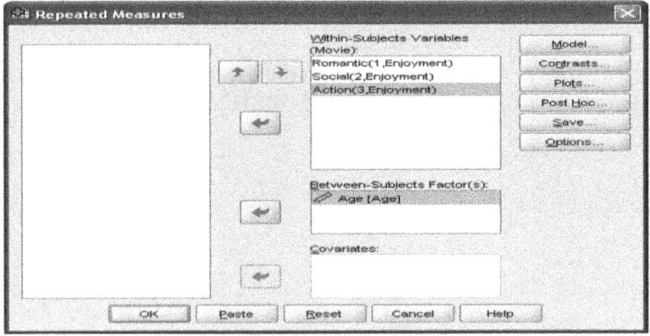

Figure 6.8 Screen showing option for selecting within-subjects (movie) and between-subjects variables (Age)

and action movie) from the left panel and bring them in the "Within-Subjects Variables" section by clicking on the arrow key. Similarly bring between-subjects factor 'Age' into "Between-Subjects Factor(s)" section.

Click on **Options** command in Figure 6.8 to open the Repeated Measures Options sub-dialogue box as shown in Figure 6.9. In this screen transfer the factors Age, Movie, and Age × Movie from the "Factor(s) and Factor Interactions" section in the left panel to the "Display Means for" section in the right panel by using arrow button. This will generate marginal means of Movie and means for the data in each cell.

Check the option 'Compare main effects' and select 'Bonferroni' option in the "Confidence Interval adjustment" dialogue box. This will generate outputs for testing the main effects of within-subjects factor (Movie). There is no post-hoc test available for within-subjects factor; hence Bonferroni correction is required to be selected to

ILLUSTRATION   **139**

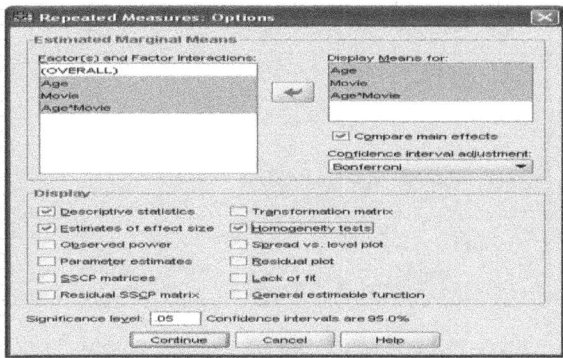

Figure 6.9 Screen showing options for comparing main effect for within-subjects factor (Movie) and other statistics

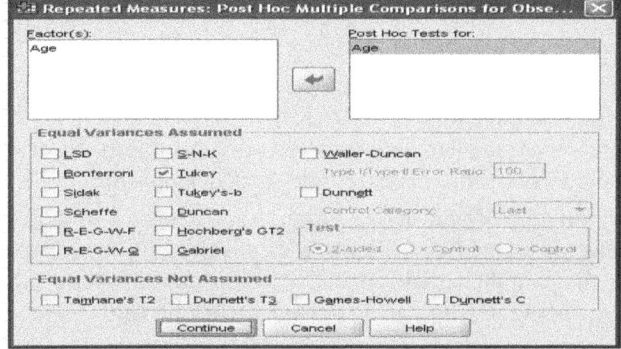

Figure 6.10   Screen showing option for post-hoc test for the between-subjects factor (Age)

compensate the inflated type I error due to multiple comparisons. For detail discussion on this correction, readers are advised to refer to Chapter 3.

Check 'Descriptive statistics', 'Estimates of effect size', and 'Homogeneity tests' options in the "Display" area for generating various outputs related to descriptive statistics, efficiency, and homogeneity condition. Click on **Continue** to get back to the screen shown in Figure 6.8.

Click on **Post-hoc** option in the screen shown in Figure 6.8 for testing the main effect of between-subjects factor (Age). Transfer the variable *Age* from the "Factor(s)" section to the "Post-hoc Tests" section in the screen as shown in Figure 6.10. Select 'Tukey' as an option for post-hoc test. This is the most appropriate option for the post-hoc test if the assumptions are satisfied. Click on **Continue** for further options. This will take you back to the screen shown in Figure 6.8.

In Figure 6.8 click on **Plots** command for generating different means plots. Do the following:

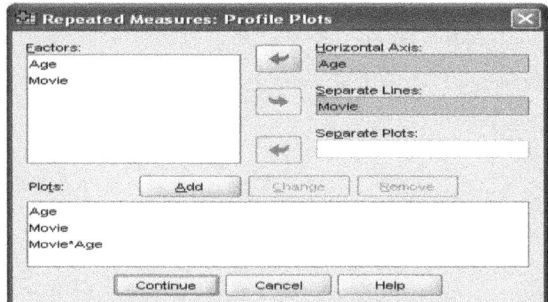

Figure 6.11   Screen showing option for means plots

a. Transfer the variable *Age* from the "Factors" section to the "Horizontal Axis" area for generating means plot to compare the main effect of between-subjects factor (Age) on enjoyment. Click on **Add**.

b. Transfer the variables *Movie* from the "Factors" section to the "Horizontal Axis" area for generating means plot to compare the main effect of within-subjects factor (Movie) on enjoyment. Click on **Add**.

c. Transfer variables *Age* and *Movie* from the "Factors" section to the "Separate Lines" and "Horizontal Axis" areas, respectively, for generating means plots in order to compare the simple effects of between-subjects factor (Age) in each level of the within-subjects factor (Movie). Click on **Add**.

d. Transfer variables *Movie* and *Age* from the "Factors" section to the "Separate Lines" and "Horizontal Axis" areas, respectively, for generating means plots in order to compare the simple effects of within-subjects factor (Movie) in each level of the between-subjects factor (Age). Click on **Add** (Figure 6.11).

After selecting the above-mentioned options for the means plots click on **Continue**. This will take you back to the screen shown in Figure 6.8. Click on **OK** for generating outputs in this design.

SPSS Outputs and Interpretation

In mixed design analysis SPSS generates lots of output in the output window, but only few outputs as mentioned below need to be selected for interpretation.

1. Box's test for homogeneity of variance–covariance matrices (Table 6.3)
2. Levene's test for equality of error variances (Table 6.4)
3. Mauchly's Test of sphericity (Table 6.5)
4. Descriptive statistics (Table 6.6)
5. *F*-Table for testing significance of Movie (within-subjects) effects and Inter-action effect (Movie × Age) (Table 6.7)
6. Pair-wise comparison of marginal means of Movie (within-subjects) (Table 6.8)

ILLUSTRATION   141

7. *F*-table for testing significance of Age (between-subjects) effects (Table 6.9)
8. Pair-wise comparison of marginal means of Age (between-subjects) (Table 6.10)
9. Marginal means plot of Movie (Figure 6.12)
10. Marginal means plot of Age (Figure 6.13)
11. Marginal means plot of Age × Movie (Figure 6.17)
12. Marginal means plot of Movie × Age (Figure 6.21)

The output in Table 6.7 is used to test the main effect of within-subjects factor (Movie) and the interaction between Movie × Age, whereas the output in Table 6.9 facilitates us to test the main effect of between-subjects factor (Age). If the interaction effect is significant, investigating main effects becomes meaningless and in that case simple effects are investigated by using the same data file in SPSS. The procedure shall be discussed in detail in the later part of this chapter. Thus, in case interaction effect is significant, following additional outputs are generated to discuss simple effects of within-subjects and between-subjects factors:

13. Repeated measures ANOVA's for testing simple effect of Movie (within-subjects)
    a. Mauchly's test of sphericity in different age categories (Table 6.11)
    b. F-Table for testing significance of Movie (within-subjects) effect in each age category (Table 6.12)
    c. Pair-wise comparisons of means in each Age category (Table 6.13)
14. Independent measures ANOVA's for testing the simple effect of Age (between-subjects)
    d. Levene's test for equality of variances. (Table 6.14)
    e. *F*-Table for testing significance of Age (between-subjects) effect in each movie category (Table 6.15)
    f. Pair-wise comparisons of means in each Movie category (Table 6.16)

Testing Assumptions

Before analyzing the findings let us first check the assumptions required for using the mixed design.

Assumption of Normality   This assumption can be tested for the data on enjoyment in each of the nine cells with the help of SPSS by using the procedure described earlier in the Chapter 3. The normality is tested by means of Shapiro–Wilk statistic. For the data to be normal, the Shapiro–Wilk statistic should not be significant, that is, *p* value should be ≥ 0.05. Since this statistic is not significant in any of the nine cells as shown in the Table 6.2, the assumption of normality is not violated.

Table 6.2   Test of Normality

		Shapiro–Wilk Statistic	df	Sig. (p-Value)
Teens	Romantic	0.893	6	0.332
	Social	0.949	6	0.729
	Action	0.920	6	0.509
Mid age	Romantic	0.964	6	0.847
	Social	0.892	6	0.327
	Action	0.929	6	0.574
Old age	Romantic	0.911	6	0.443
	Social	0.979	6	0.946
	Action	0.990	6	0.988

Table 6.3   Box's Test for Equality of Covariance Matrices

Box's M	9.453
F	0.552
df1	12
df2	1.090E3
Sig. (p-value)	0.881

Tests the null hypothesis that the observed covariance matrices of the dependent variables are equal across groups.

Homogeneity of Variance Covariance Matrices   One of the assumptions of the mixed design is that the pattern of inter-correlations among the various levels of the within-subjects factor should be similar in different levels of the between-subjects factor. This assumption holds true if the Box's M test is non-significant. It can be seen in the Table 6.3 that the Box's M test is non-significant; hence the assumption of homogeneity of variance–covariance matrices holds true. Since this assumption is very sensitive in nature, the level of significance for Box's M test should be taken as 0.001. In other words, you can consider this test to be non-significant only if the p value is equal or more than 0.001.

Homogeneity of Variance   This assumption is required to be tested because the mixed design consists of between-subjects factor also. This is an assumption of the independent measures ANOVA. SPSS uses Levene's test for this assumption. Since Levene's statistic for each level of the within-subjects factor (Movie) is non-significant ($p \geq 0.05$) as shown in the Table 6.4, the homogeneity of variance assumption has not been violated.

Sphericity Assumption   The sphericity assumption is required to be tested in mixed design because it has a within-subjects factor. In SPSS sphericity assumption is tested by means of Mauchly's W test. For sphericity assumption to be true, the Mauchly's test should be non-significant. Table 6.5 shows that the Mauchly's W test is non-significant ($p > 0.05$), hence the sphericity assumption is not violated.

ILLUSTRATION   **143**

Table 6.4   Levene's Test for Equality of Error Variances

	Levene Statistic	df1	df2	Sig. (p-Value)
Romantic	0.188	2	15	0.830
Social	2.532	2	15	0.113
Action	0.422	2	15	0.663

Table 6.5   Mauchly's Test of Sphericity

Measure: Enjoyment

Within Subjects Effect	Mauchly's W	Approx. Chi-Square	df	Sig. (p-Value)	Epsilon		
					Greenhouse– Geisser	Huynh– Feldt	Lower- bound
Movie	0.888	1.663	2	0.435	0.899	1.000	0.500

Table 6.6   Descriptive Statistics

	Age Category	Mean	SD	N
Romantic movie	Teens	64.0000	2.96648	6
	Mid_age	61.1667	2.63944	6
	Old_age	59.3333	2.50333	6
	Total	61.5000	3.22217	18
Social movie	Teens	50.8333	3.25064	6
	Mid_age	60.8333	5.98052	6
	Old_age	66.6667	3.32666	6
	Total	59.4444	7.88313	18
Action movie	Teens	58.1667	4.16733	6
	Mid_age	46.8333	3.54495	6
	Old_age	50.8333	3.60093	6
	Total	51.9444	5.99482	18

Since sphericity assumption is not violated, no correction is required and the significance value of p associated with the F for the 'Sphericity Assumed' case mentioned in the Table 6.7 ($p = 0.000$) shall be reported in the findings.

Descriptive Statistics

Table 6.6 refers to the descriptive statistics of the enjoyment scores of the subjects in each age category tested in all the movie conditions. These results can be used to compare the mean and standard deviation in each cell. These means are used to prepare means plots for comparing main and simple effects.

Table 6.7   *F*-Table for Testing Significance of Movie (Within-Subjects) Effects

Measure:Enjoyment

Source		Type III SS	df	Mean Square	F	Sig. (p-Value)	Partial Eta Squared
Movie	**SphericityAssumed**	910.704	2	455.352	**54.545**	**0.000**	0.784
	Greenhouse–Geisser	910.704	1.799	506.360	54.545	0.000	0.784
	Huynh–Feldt	910.704	2.000	455.352	54.545	0.000	0.784
	Lower-bound	910.704	1.000	910.704	54.545	0.000	0.784
Movie* Age	**Sphericity Assumed**	1168.185	4	292.046	**34.983**	**0.000**	0.823
	Greenhouse–Geisser	1168.185	3.597	324.761	34.983	0.000	0.823
	Huynh–Feldt	1168.185	4.000	292.046	34.983	0.000	0.823
	Lower-bound	1168.185	2.000	584.093	34.983	0.000	0.823
Error (Movie)	Sphericity Assumed	250.444	30	8.348			
	Greenhouse–Geisser	250.444	26.978	9.283			
	Huynh–Feldt	250.444	30.000	8.348			
	Lower-bound	250.444	15.000	16.696			

Bold face indicates that the effect is significant at 5% level.

Testing Main Effect of Movie (within-Subjects)

In this factorial design there are two independent factors, Movie and Age, whose effects need to be investigated. Here the Movie is a within-subjects factor and Age is a between-subjects factor. In this illustration interaction effect is significant, hence analyzing the main effects becomes meaningless. However, these analyses have been shown for explaining the procedure involved in it.

Since sphericity assumption has not been violated, the *F* value for Movie and its corresponding significance value (*p*-value) mentioned in front of the 'Sphericity Assumed' in Table 6.7 shall be considered for the interpretation. Since *p*-value corresponding to *F* for Movie is less than 0.05 and partial eta square is 0.784, the main effect of Movie (within subjects) is meaningful and significant at 5% level and null hypothesis that the average enjoyment scores is same in all the three movie groups irrespective of age categories is rejected. Since the main effect of Movie is significant, pair-wise comparison among different movie groups irrespective of the Age shall be done.

Pair-Wise Comparison of Marginal Means of Movie Groups   Since post-hoc test cannot be used in repeated measures design, no option is available in SPSS for post-hoc analysis for the within-subjects variable (Movie). SPSS uses paired t tests to compare group means of within-subjects factor. Due to multiple comparisons, the family-wise error rate (α) inflates and to compensate this error Bonferroni Correction (for detail see Chapter 2) was used while comparing (Figure 6.9). SPSS uses Bonferroni correction for each comparison of paired group means and provides p-value for testing the significance of mean difference as shown in Table 6.8.

It can be seen from the Table 6.8 that there is a significant difference between the enjoyment scores of Romantic and Action movie groups as well as between social and action movie groups because the *p* value associated with these differences is 0.000

ILLUSTRATION   145

Table 6.8   Pair-Wise Comparison of Marginal Means of Movie Irrespective of Age

Measure: Enjoyment

| Mean Difference | | | | | 95% CI for Difference[a] | |
Movie (*I*)	Movie (*J*)	(*I–J*)	Std. Error	Sig.[a] (*p*-value)	Lower Bound	Upper Bound
Romantic	social	2.056	0.898	0.111	−0.362	4.473
	Action	9.556*	1.112	**0.000**	6.561	12.550
Social	Romantic	−2.056	0.898	0.111	−4.473	0.362
	Action	7.500*	0.861	**0.000**	5.181	9.819
Action	Romantic	−9.556*	1.112	**0.000**	−12.550	−6.561
	Social	−7.500*	0.861	**0.000**	−9.819	−5.181

Based on estimated marginal means
[a] Adjustment for multiple comparisons: Bonferroni.
*The mean difference is significant at the 0.05 level.
Bold face indicates that the effect is significant at 5% level.

which is less than 0.05. Pair-wise comparison of marginal means has been shown graphically in Figure 6.12 by using the information listed in the Tables 6.6 and 6.8.

Means Plot of Movie   Figure 6.12 shows the means plot of the movie groups. It can be noted from this figure that the subjects enjoyed romantic as well as social movies more than action movies, irrespective of their age categories. Watching romantic and social movies was equally enjoyable, whereas action movie was least enjoyed by the subjects.

Testing Main Effect of Age (between-Subjects)

In this illustration Age is a between-subjects factor; hence the assumption of equality of variance covariance matrix in each of the age category needs to be satisfied.

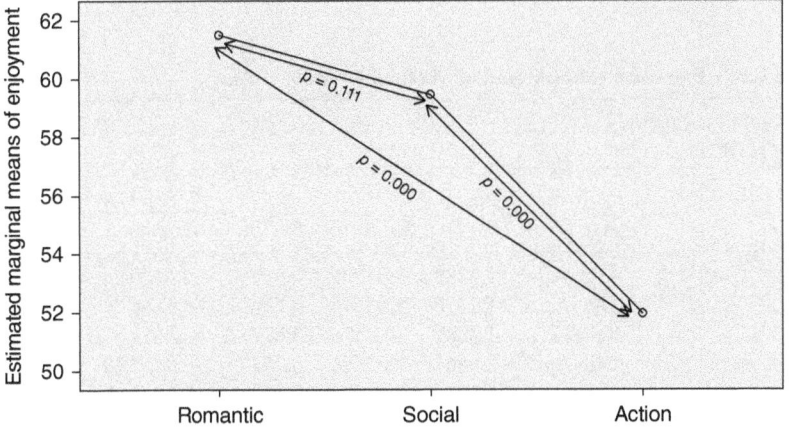

Figure 6.12   Marginal means plot of Movie

This assumption was tested by means of the Box's M test shown in Table 6.3. Since Box's M test is not significant, hence this assumptions holds and the F value can be computed for testing the significance of the between-subjects effect (Age).

Table 6.9 shows that the F value for the between-subjects factor, Age, is not significant as its corresponding p value is 0.294 which is more than 0.05. Thus, the null hypothesis of no difference of mean enjoyment scores among different age categories irrespective of the movie is not rejected on the basis of the observed sample. Since F is not significant, no post-hoc test is required in this case. However, to make the readers understand about the procedure involved in post-hoc test, this analysis has been shown along with means plot.

Pair-Wise Comparison of Marginal Means of Age Groups The readers may use the Tukey HSD post-hoc test for the pair-wise comparison among different groups of between-subjects factor. Since F value for the Age is not significant, none of the group mean difference as shown in the Table 6.10 is significant.

Means Plot of Age Figure 6.13 is a means plot of the age groups obtained in the SPSS output. It clearly indicates that none of the p-values are significant ($p > 0.05$), hence enjoyment levels among the three different age groups do not differ irrespective of the movie type.

Table 6.9   *F*-table for testing significance of Age (between-subjects) effects

Measure:MEASURE_1
Transformed Variable:Average

Source	Type III	df	Mean Square	F	Sig. (p-Value)	Partial Eta Squared
Intercept	179 343.407	1	179 343.407	7.447E3	0.000	0.998
Age	64.037	2	32.019	1.330	0.294	0.151
Error	361.222	15	24.081			

Table 6.10   Pair-wise Comparison of Marginal Means of Age

Measure: Enjoyment
Tukey HSD

Mean Difference						95% CI for Difference	
Age (I)	Age (J)		($I - J$)	Std. Error	Sig*	Lower Bound	Upper Bound
Teens	Mid_Age		1.3889	1.63576	0.679	−2.8600	5.6377
	Old_Age		−1.2778	1.63576	0.720	−5.5266	2.9711
Mid_Age	Teens		−1.3889	1.63576	0.679	−5.6377	2.8600
	Old_Age		−2.6667	1.63576	0.264	−6.9155	1.5822
Old_Age	Teens		1.2778	1.63576	0.720	−2.9711	5.5266
	Mid_Age		2.6667	1.63576	0.264	−1.5822	6.9155

* Significant at 5% level.

ILLUSTRATION   147

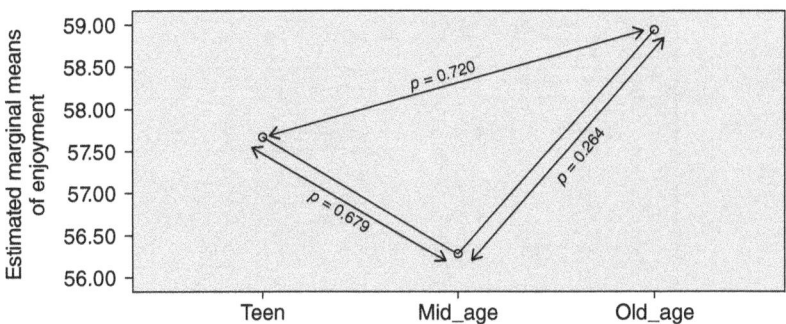

Figure 6.13   Marginal means plot of Age

Testing Significance of Interaction (Movie × Age)

Table 6.7 indicates that the F value for the interaction (Movie*Age) is significant and meaningful because the p value associated with the F (Sphericity Assumed) is 0.000, which is less than even 0.05 and partial eta square is 0.823 which is considered to be very high. In other words, the null hypothesis of no interaction between movie and age is rejected at 5% level. Since sphericity assumption has not been violated, no correction has been made to the degrees of freedom and the p value against 'Sphericity Assumed' row has been considered for testing the significance of F value. Since the interaction is significant, it is important to investigate the simple effects of Movie as well as Age.

Simple Effect of Movie (within-Subjects)   To find the simple effect of Movie, we shall compare the enjoyment scores of all the three movie groups in each age category separately. In other words, the null hypothesis of no difference between the enjoyment scores in all the three movie groups shall be tested separately in each age category. To do so three separate one-way repeated measures ANOVA need to be applied. This can be done by splitting the data file in SPSS by using the following sequence of commands:

Data ⟶ Split File

After clicking on the **Split File** option the screen shown in Figure 6.14 shall be obtained. Choose the radio button 'Organize output by groups' and bring the variable *Age* category from left panel into the area marked with "Grouped based on" in the right panel. Ensure that the radio button 'Sort the file by grouping variables' is selected. This option is in fact selected by default. Click on **OK** to get the data file splitted as per the age category. The SPSS will show the following message in the output dialogue box:

SORT CASES BY Age.
SPLIT FILE SEPARATE BY Age

Go back to the data file and follow the commands for repeated measures designs as narrated while referring to Figures 6.6 to 6.7. After following the commands for

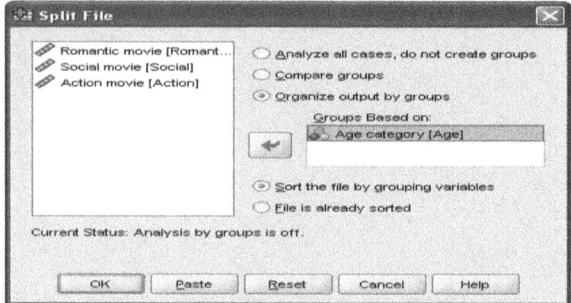

Figure 6.14   Screen showing option for splitting the data file for one-way repeated measures ANOVA

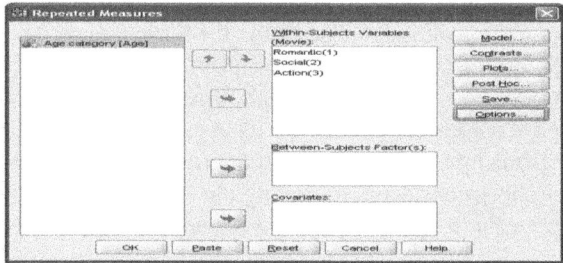

Figure 6.15   Screen showing option for selecting within-subjects variables

one-way repeated measures ANOVA (define within-subject factor name as Movie with 3 levels and measure name as Enjoyment), you will get the screen as shown in Figure 6.15 to select the repeated measures variables. Select all the three variables namely, *Romantic, Social*, and *Action* and drag them into the "Within-Subjects Variables" area on the right side of the screen. Click on **OK** to get the screen as shown in Figure 6.16.

In the screen shown in Figure 6.16, transfer the variable *Movie* from the left panel to the "Display Means for" section in the right side of the screen. Check the option 'Compare main effects' and select 'Bonferroni' correction as shown in the screen. Further, select the options for 'Descriptive statistics', 'Estimates of effect size' to generate the relevant findings in investigating the simple effects. Click on **Continue** to generate the outputs as shown in Tables 6.11–6.13. The SPSS outputs for all the three levels of the Movie (within-subjects) have been combined in these tables.

Since different levels of the Movie are within-subjects factors; hence before comparing the effect of different movie in each age category, it is necessary to test the assumption of sphericity. Table 6.11 shows that the sphericity assumption is not violated in all the three age categories as none of the p-value is less than 0.05.

ILLUSTRATION 149

Figure 6.16 Screen showing option for pair-wise comparison of group means using Bonferroni correction

Table 6.11 Mauchly's test of sphericity[a] in different age categories

Measure: Enjoyment

Age Category	Within Subjects Effect	Mauchly's W	Approx. Chi-Square	df	Sig. (p-value)	Epsilon[b]		
						Greenhouse–Geisser	Huynh–Feldt	Lower-bound
Teens	Movie	0.392	3.746	2	0.154	0.622	0.727	0.500
Mid Age	Movie	0.843	0.684	2	0.710	0.864	1.000	0.500
Old Age	Movie	0.635	1.814	2	0.404	0.733	0.961	0.500

Tests the null hypothesis that the error covariance matrix of the orthonormalized transformed dependent variables is proportional to an identity matrix.
[a]Design: Intercept + Age
[b]May be used to adjust the degrees of freedom for the averaged tests of significance. Corrected tests are displayed in the Tests of Within-Subjects Effects table.
Within Subjects Design: Movie

Since sphericity assumption has not been violated, the value of F and its associated significance value (p) for 'Sphericity Assumed' case would be considered for interpreting the findings in all three age categories. Table 6.12 shows that the F values for Movie in Teens, Mid-Age, and Old-Age are significant as their associated p values are less than 0.017. Thus, the null hypothesis of no difference in mean enjoyment scores among the three movie groups is rejected in each age category. Since the effect of Movie is significant in each age category, it is important to do the pair-wise comparison of means among three levels of movie in each age category which is shown in Table 6.13.

Pair-wise Comparison of Means in Each Age Category Pair-wise comparison of means among three levels of movie in each age category has been shown in Table 6.13. Depending upon the level of significance set by the researcher, the mean difference can be tested for its significance by looking to its associated p-value. Since in repeated

Table 6.12 F-Table for Testing Significance of Movie (Within-Subjects) Effects in Each Age Category

Measure: **Enjoyment**

Source		Type III SS	df	Mean Square	F	Sig. (p-value)	Partial Eta Squared
Teens							
Movie	**Sphericity Assumed**	522.333	2	261.167	**41.898**	**0.000**	0.893
	Greenhouse–Geisser	522.333	1.244	419.955	41.898	0.000	0.893
	Huynh–Feldt	522.333	1.454	359.162	41.898	0.000	0.893
	Lower-bound	522.333	1.000	522.333	41.898	0.001	0.893
Error							
(Movie)	Sphericity Assumed	62.333	10	6.233			
	Greenhouse–Geisser	62.333	6.219	10.023			
	Huynh–Feldt	62.333	7.272	8.572			
	Lower-bound	62.333	5.000	12.467			
Mid-Age							
Movie	**Sphericity Assumed**	803.111	2	401.556	**42.618**	**0.000**	0.895
	Greenhouse–Geisser	803.111	1.728	464.660	42.618	0.000	0.895
	Huynh–Feldt	803.111	2.000	401.556	42.618	0.000	0.895
	Lower-bound	803.111	1.000	803.111	42.618	0.001	0.895
Error (Movie)	Sphericity Assumed	94.222	10	9.422			
	Greenhouse–Geisser	94.222	8.642	10.903			
	Huynh–Feldt	94.222	10.000	9.422			
	Lower-bound	94.222	5.000	18.844			
Old-Age							
Movie	**Sphericity Assumed**	753.444	2	376.722	**40.124**	**0.000**	0.889
	Greenhouse–Geisser	753.444	1.466	514.047	40.124	0.000	0.889
	Huynh–Feldt	753.444	1.922	391.933	40.124	0.000	0.889
	Lower-bound	753.444	1.000	753.444	40.124	0.001	0.889
Error (Movie)	Sphericity Assumed	93.889	10	9.389			
	Greenhouse–Geisser	93.889	7.329	12.811			
	Huynh–Feldt	93.889	9.612	9.768			
	Lower-bound	93.889	5.000	18.778			

Bold face indicates that the effect is significant at 5% level.

measures design there is no post-hoc test available and multiple comparisons inflate the family wise error rate (α), Bonferroni correction has been used for pair-wise comparison by applying the paired t test. All these should not bother the researcher as this is taken care of by the SPSS once the Bonferroni correction is chosen in the steps discussed above. From the results of pair-wise comparison the following conclusion can be drawn:

In *Teens category* there is a significant difference ($p < 0.017$) between romantic and social movies as well as between social and action movies ($p < 0.017$).

In *Mid age* category there is a significant difference between romantic and action movies ($p < 0.017$) as well as social and action movies ($p < 0.017$).

In *Old age* category there is a significant difference between romantic and social movies ($p < 0.017$) as well as social and action movies ($p < 0.017$).

ILLUSTRATION   151

Table 6.13   Pair Wise Comparisons of Mean *s* in Each Age Category

Measure: Enjoyment

Movie (*I*)	Movie (*J*)	Mean Difference (*I–J*)	Std. Error	Sig[a]	95% CI for Difference[a] Lower Bound	Upper Bound
Age_Category = Teens						
Romantic	Social	13.167*	1.167	**0.000**	9.044	17.290
	Action	5.833	1.922	0.087	−0.960	12.626
Social	Romantic	−13.167*	1.167	**0.000**	−17.290	−9.044
	Action	−7.333*	1.085	**0.003**	−11.169	−3.498
Action	Romantic	−5.833	1.922	0.087	−12.626	0.960
	Social	7.333*	1.085	**0.003**	3.498	11.169
Age_Category = Mid_Age						
Romantic	Social	0.333	2.092	1.000	−7.061	7.728
	Action	14.333*	1.626	**0.001**	8.586	20.080
Social	Romantic	−0.333	2.092	1.000	−7.728	7.061
	Action	14.000*	1.549	**0.001**	8.525	19.475
Action	Romantic	−14.333*	1.626	**0.001**	−20.080	−8.586
	Social	−14.000*	1.549	**0.001**	−19.475	−8.525
Age_Category = Old_Age						
Romantic	Social	−7.333*	1.229	**0.006**	−11.678	−2.989
	Action	8.500	2.187	0.035	0.771	16.229
Social	Romantic	7.333*	1.229	**0.006**	2.989	11.678
	Action	15.833*	1.759	**0.001**	9.616	22.050
Action	Romantic	−8.500	2.187	0.035	−16.229	−0.771
	Social	−15.833*	1.759	**0.001**	−22.050	−9.616

Based on estimated marginal means
*The mean difference is significant at the 0.017 level.
[a] Adjustment for multiple comparisons: Bonferroni.
Bold face indicates that the effect is significant at 5% level.

In order to know, watching which movie, enjoyment is more in different age categories a means plot has been shown in Figure 6.17. This has been obtained in the main output of the SPSS during the analysis of the main effects.

Means Plot (Age × Movie)   Post-hoc test results of the simple effect analysis shown in the Table 6.13 and the contents of Table 6.2 can be used to show the means plots of different movies group in each age category as shown in Figure 6.17. This means plot provides clear picture about the analysis. It indicates that the enjoyment in watching romantic movie as well as action movie is significantly higher than that of the social movies in the teen category as the *p* values are less than 0.017. In the Mid age category the trend is different. In this category enjoyment of the subjects is more in social as well as romantic movie in comparison to that of action movie. However, the subjects in the Old age category found social movie to be the most enjoyable in comparison to that of romantic as well as action movie.

Simple Effect of Age (between-Subjects)   To find the simple effect of Age it is required to compare the enjoyment scores of all the three age groups in each movie

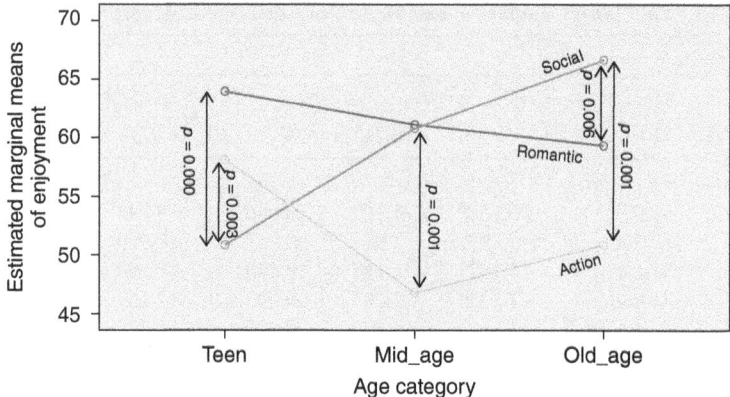

Figure 6.17   Marginal means plot of Age × Movie

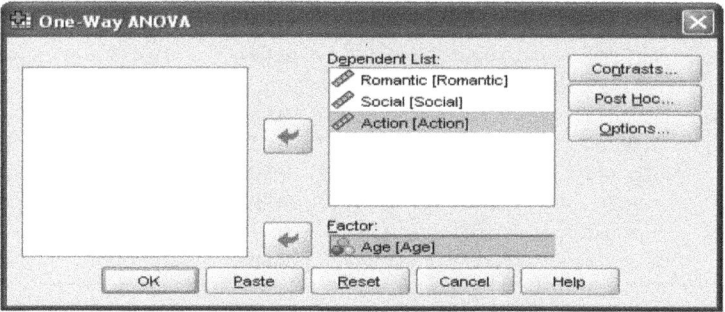

Figure 6.18   Screen showing option for selecting the variables for analyzing simple effect of between-subjects variable

category separately. In other words, three separate one-way independent measures ANOVA need to be applied. This can be done by using the same data file created originally (without splitting) and using the following sequence of commands in SPSS:

Analyze ⟶ Descriptive statistics ⟶ Explore

After using this sequence of commands Figure 6.18 is obtained. Bring all three variables, *Romantic, Social*, and *Action*, from the left panel into the "Dependent List" section of the right panel. Transfer Age_Category variable from left panel to the "Factor" section on the right panel of the window. Click on the **Post-hoc** tag and select 'Tukey' option as shown in Figure 6.19. Click on **Continue** to go back to Screen shown in Figure 6.18. Now click on the **Option** tag and select the option for 'Descriptive' and 'Homogeneity of variance test' as shown in Figure 6.20 and click on **Continue**. Click on **OK** to get the results as shown in the Tables 6.14–6.16.

Since homogeneity of variance assumption has not been violated in all the three movie groups as shown in Table 6.14, ANOVA for independent measures can be

ILLUSTRATION   153

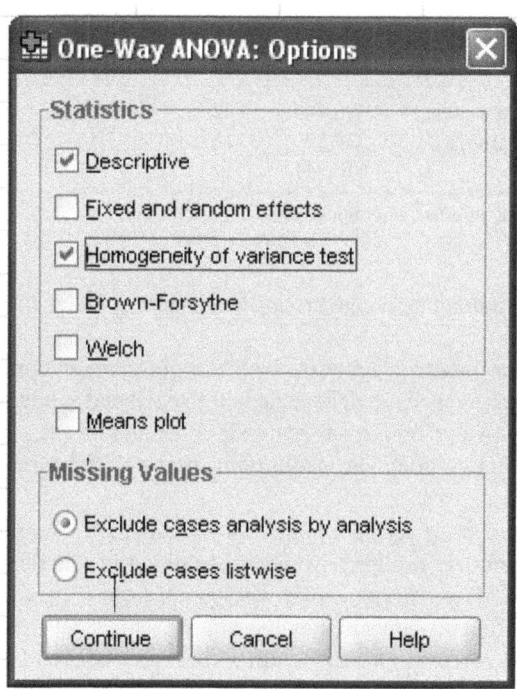

Figure 6.19   Screen showing option for post-hoc test

Figure 6.20   Screen showing option for descriptive statistics and testing assumption

applied for testing the null hypothesis of equal mean enjoyment scores in each movie category separately.

Table 6.15 indicates that the value of F for the Social movie ($p = 0.000$) and Action movie ($p = 0.000$) is significant as the p values associated with them are less than 0.017. Hence, it can be concluded that the null hypothesis of no difference in mean

Table 6.14   Test of Homogeneity of Variances

	Levene Statistic	df1	df2	Sig.
Romantic	0.188	2	15	0.830
Social	2.532	2	15	0.113
Action	0.422	2	15	0.663

Table 6.15   *F*-Table for Testing Significance of Age (between-Subjects) Effects in Each Movie Category

Movie		Sum of Squares	df	Mean Square	*F*	Sig. (*p*-Value)
Romantic	Between groups	66.333	2	33.167	4.516	0.029
	Within groups	110.167	15	7.344		
	Total	176.500	17			
Social	Between groups	769.444	2	384.722	20.107	**0.000**
	Within groups	287.000	15	19.133		
	Total	1056.444	17			
Action	Between groups	396.444	2	198.222	13.862	**0.000**
	Within groups	214.500	15	14.300		
	Total	610.944	17			

Bold face indicates that the effect is significant at 5% level.

enjoyment scores among three age groups is rejected in social and action movie categories.

To investigate as in which age group subjects get more enjoyment in each movie category, a post-hoc analysis shall be done in the social and action movie by using the Tukey test. Since *F* value for Romantic movie is not significant, no post-hoc analysis shall be done for this variable. The results of the post-hoc test are shown in Table 6.16.

Pair-wise Comparison of Means in each Movie Category   Results of the post-hoc tests for all three movie groups have been combined in the Table 6.16. Following conclusions can be drawn from the pair-wise comparison of means at 5% level in the post-hoc analysis outputs generated by the SPSS.

In *Social movie category* enjoyment differs between teens and mid age subjects ($p < 0.017$) as well as between teens and old age subjects ($p < 0.017$), whereas mid and old age subjects do not differ in their enjoyment level.

In Action movie subjects in the teens and mid age category ($p < 0.017$) as well as teens and old age category ($p < 0.017$) differ significantly in their mean enjoyment scores. However, enjoyment does not differ between mid age and old age category.

In order to know as to which age category subjects enjoy more during watching different movies, means plot has been shown in Figure 6.21. This has been obtained in the main output of the SPSS during analysis of the main effects.

ILLUSTRATION   155

Table 6.16   Pair-Wise Comparisons of Mean s in Each Movie Categories

Correction: Tukey HSD

Dependent Variable	Age Category (I)	Age Category (J)	Mean Diff. (I–J)	Std. Error	Sig. (p-Value)	95% Confidence Interval	
						Lower Bound	Upper Bound
Romantic	Teens	Mid age	2.83333	1.56466	0.200	−1.2308	6.8975
		Old age	4.66667	1.56466	0.024	0.6025	8.7308
	Mid age	Teens	−2.83333	1.56466	0.200	−6.8975	1.2308
		Old age	1.83333	1.56466	0.487	−2.2308	5.8975
	Old age	Teens	−4.66667	1.56466	0.024	−8.7308	−0.6025
		Mid age	−1.83333	1.56466	0.487	−5.8975	2.2308
Social	Teens	Mid age	−10.00000*	2.52543	**0.003**	−16.5597	−3.4403
		Old age	−15.83333*	2.52543	**0.000**	−22.3931	−9.2736
	Mid age	Teens	10.00000*	2.52543	**0.003**	3.4403	16.5597
		Old age	−5.83333	2.52543	0.085	−12.3931	0.7264
	Old age	Teens	15.83333*	2.52543	**0.000**	9.2736	22.3931
		Mid age	5.83333	2.52543	0.085	−0.7264	12.3931
Action	Teens	Mid age	11.33333*	2.18327	**0.000**	5.6624	17.0043
		Old age	7.33333*	2.18327	**0.011**	1.6624	13.0043
	Mid age	Teens	−11.33333*	2.18327	**0.000**	−17.0043	−5.6624
		Old age	−4.00000	2.18327	0.193	−9.6710	1.6710
	Old age	Teens	−7.33333*	2.18327	**0.011**	−13.0043	−1.6624
		Mid age	4.00000	2.18327	0.193	−1.6710	9.6710

*The mean difference is significant at the 0.017 level.
Bold face indicates that the effect is significant at 5% level.

Means Plot (Movie × Age)   The means plot shown in the Figure 6.21 has been generated by using the contents of Table 6.2 and 6.16. This figure shows the means plots of different age category in each movie group. It gives clear picture about the analysis. It can be seen from this plot that social movie is enjoyed maximum by the old age subjects and least by the teens. The story is entirely different while watching action movie. In such movie teens derive maximum enjoyment, whereas mid age subjects get the least.

How to Report the Findings

In this study a mixed design analysis was used to test the significance of main and simple effects of Movie (between-subjects) and Age (within-subjects) factors. The hypotheses were tested at the significance level 0.05.

Assumptions   In order to test the suitability of using the mixed design, its required assumptions were tested which are reported as follows:

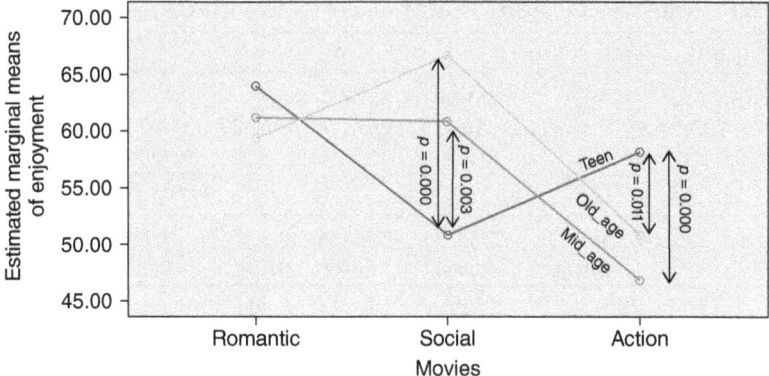

Figure 6.21   Marginal means plot of Movie × Age

→ Since Shapiro–Wilk statistic was not significant in each of the nine group of scores, assumption of normality was not violated (Table 6.2)

→ The Box's M test was non-significant, hence the assumption of homogeneity of variance–covariance matrices was satisfied (Table 6.3)

→ Levene's statistic was non-significant in each movie group, hence the assumption of homogeneity of variance was satisfied (Table 6.4)

→ Mauchly's W test was non-significant, hence the sphericity assumption was satisfied and no correction was required in the degrees of freedom for testing the significance of main effect of Movie (within-subjects) factor (Table 6.5)

Testing Main Effects

→ Since F value for Movie was significant, the null hypothesis that the average enjoyment score was same in all the three movie groups irrespective of age categories was rejected at 5% level. The pair-wise comparison among different movie groups irrespective of the age was done by using the Bonferroni correction. The results indicated that the subjects enjoyed romantic as well as social movies more than action movies irrespective of their age categories (Tables 6.7 and 6.8, Figure 6.12)

→ Since F value for Age was not significant, the null hypothesis of no difference of mean enjoyment scores among different age categories irrespective of the movie was not rejected on the basis of the observed sample (Table 6.9)

Testing Simple Effects

→ Since F value for the interaction (Movie*Age) was significant, interaction existed between Movie and Age (Table 6.7) and therefore simple effects were investigated.

→ The value of F in Teens, Mid-Age, and Old-Age was significant; hence the null hypothesis of no difference in mean enjoyment scores among the three movie groups was rejected at the significance level 0.05 in each age category (Tables 6.12 and 6.13, Figure 6.17). The pair-wise analysis suggests the following:

a. Teens enjoy more watching romantic and action movies in comparison to social movie.

b. Mid age subjects enjoy more watching social and romantic movies rather than action movie.

c. In old age category subject's enjoyment is more in watching social movie and least in action movie.

→ Since the value of F for Social and Action movies is significant at 5% level, it may be concluded that the null hypothesis of no difference in mean enjoyment scores among the three age groups is rejected at 5% level in social and action movie categories (Tables 6.15 and 6.16, Figure 6.21). The following may be concluded from the pair-wise comparison of means:

a. Old age subjects enjoyed more watching social movie, whereas teen's enjoyment was least.

b. Action movie was mostly liked by the teens, whereas mid age subject's enjoyment was least in that movie.

Inference

On the basis of the subjects selected in the study, it may be concluded that the teens enjoy watching romantic and action movie, mid age subjects get more fun while watching social and romantic movies, whereas old age subjects prefer social movie the most. Further, old age and mid age subjects derive maximum pleasure in watching social movie, whereas teens derive maximum pleasure in watching action movie.

EXERCISE

6.1. Explain the difference in a two-way mixed design and two-way repeated measures design. How they are different in assumptions and layout?

6.2. Explain a situation where you would like to use a two-way mixed design to address your research questions.

6.3. In a two-way mixed design, between-subjects factor has two and within-subjects factor has three levels, whereas nine subjects have been used in a study. Explain testing protocol by showing layout design in the study.

6.4. Write in brief the steps involved in solving a two-way mixed design using SPSS.

6.5. In a two-way mixed design if between-subjects factor and within-subjects factor have two and four levels, respectively, whereas eight subjects in the sample have been tested, show how the total sum of squares would be distributed into different components along with degrees of freedom.

ASSIGNMENT

6.1. An exercise scientist wanted to investigate the effect of gender and duration on weight reduction of participants in a fitness intervention program. A randomly selected sample of 20 subjects was given a fitness intervention program for six weeks. The sample consisted of 10 male and 10 female subjects. Weights of these subjects were measured on zero day, three weeks, and six weeks which are shown in Table 6.17. Solve this mixed design and test the main and simple effects by assuming family-wise error rate as 0.05.

Table 6.17   Weights of the Subjects Measured at Different Duration During Exercise Intervention

	Subject	Zero Day	Three Weeks	Six Weeks
Male	1	65	57	50
	2	60	62	56
	3	57	53	46
	4	61	54	50
	5	62	60	52
	6	62	63	51
	7	60	58	48
	8	67	65	53
	9	48	59	44
	10	55	52	43
Female	1	72	66	46
	2	65	58	47
	3	67	65	51
	4	62	52	40
	5	68	69	46
	6	66	60	48
	7	65	55	40
	8	72	60	46
	9	84	79	58
	10	62	60	50

6.2. A marketing company desired to investigate whether advertisement with intervention of information technology is better than the traditional method of advertising in promoting the sale of its new fruit drinks. The two advertisement campaigns, high-tech and traditional, were developed. The high-tech campaign was based on using various resources of information technology, whereas the traditional method was based on using the resources of billboard, handbills, press conference, and customer base. The cost of both the advertisements was kept same in order to have the control in the experiment. Two similar cities were chosen for the study. In one city high-tech advertisement was launched, whereas in other traditional campaign was carried out for promoting the product for three weeks. The number of units sold in each of the five outlets

Table 6.18   Sales Figure of Fruit Drink Units Sold in Different Treatment Conditions

	Outlets	T_1	T_2	T_3	T_4
High-tech adv.	1	20	25	28	32
	2	15	25	26	35
	3	22	30	35	40
	4	25	32	33	43
	5	18	28	30	40
Traditional adv.	1	16	28	29	32
	2	18	26	25	36
	3	21	30	32	38
	4	20	30	35	42
	5	17	26	27	38

T_1: First week pre-advertisement
T_2: First week post-advertisement
T_3: Second week post-advertisement
T_4: Third week post-advertisement

in both the cities was recorded in the first week prior to the advertisement and in the first, second, and third week after the launch of advertisements. The data so obtained are shown in Table 6.18. Apply two factor mixed design and discuss your findings by testing the main and simple effects. Take family-wise error rate as 0.05.

BIBLIOGRAPHY

Geisser S, Greenhouse SW. An extension of Box's result on the use of the F distribution in multivariate analysis. Ann Math Stat 1958;29:885–891.

Greenhouse SW, Geisser S. On methods in the analysis of profile data. Psychometrika 1959;24:95–112.

Gueorguieva R, Krystal JH. Progress in analyzing repeated-measures data and its reflection in papers published in the archives of general psychiatry. Arch Gen Psychiatry 2004;61:310–317.

Howell D. *Statistical Methods for Psychology*. 7th ed. Australia: Wadsworth; 2010.

Huck SW, McLean RA. Using a repeated measures ANOVA to analyze the data from a pretest-posttest design: a potentially confusing task. Psychol Bull 1975;82:511–518.

Huynh H, Feldt LS. Estimation of the Box correction for degrees of freedom from sample data in randomised block and split-plot designs. J Educ Stat 1976;1:69–82.

Hyunh H, Feldt LS. Conditions under which mean square ratios in repeated measurements designs have exact F-distributions. J Am Stat Assoc 1970;65:1582–1589.

O'Brien RG, Kaiser MK. The MANOVA approach for analyzing repeated measures designs: an extensive primer. Psychol Bull 1985;97:316–333. DOI: 10.1037/0033-2909.97.2.316.

7

ONE-WAY REPEATED MEASURES MANOVA

INTRODUCTION

In one-way repeated measures multivariate ANOVA design, effect of several levels of an independent factor on a group of dependent variables is compared. In this design subjects are repeatedly tested on more than one dependent variable under each level of the treatment factor. The multivariate-repeated measures analysis of variance is also referred as repeated MANOVA. The difference between repeated MANOVA and repeated ANOVA is only the number of dependent variables being investigated. In repeated MANOVA the effect of independent variable is seen on a group of dependent variables, whereas in repeated ANOVA the effect is investigated only on one dependent variable. In this design the independent variable can either be different treatment conditions or different time points.

If an independent factor has three or more levels, by using MANOVA one investigates whether the group difference occurs between different treatments. By using this design a researcher may investigate as to which visual time is more effective for memory retention of the object's characteristics. Consider an experiment in which randomly selected subjects are shown a number of objects, each for 2 seconds and then asked to recall the item's characteristics (name, color, and shape) as many as they can in a given time. Similarly if the experiment is repeated with 4 and 6 seconds visual time on the same subjects and their performance scores are recorded on the object's characteristics, then the researcher may identify the best visual time in recalling the overall object's characteristics.

Repeated Measures Design for Empirical Researchers, First Edition. J. P. Verma.
© 2016 John Wiley & Sons, Inc. Published 2016 by John Wiley & Sons, Inc.

In a study where time is an independent factor whose effect is to be investigated on a group of dependent variables, the focus of a researcher is to see as to how the subject's performance varies at different point of time. It is important that the time intervals should be equally spaced. This design is used in studies where improvement trend needs to be investigated. Let us consider a situation where it is desired to see the change in physiological status of subjects while undergoing an exercise programme over a period of time. A study may be planned where a random sample of subjects may be exposed to an exercise programme. The physiological status (heart rate, blood pressure, and vital capacity) of these subjects may be tested four times during the experiment that is, before an experiment, and after two, four, and six weeks. Here focus of the study is to compare physiological status of subjects at four different time points. Such studies are useful in understanding the performance trend of subjects during training. The procedure involved in solving one-way repeated measure MANOVA shall be discussed in detail by means of an illustration later in this chapter.

WHEN TO USE REPEATED MEASURES MANOVA?

Repeated measures MANOVA is preferred over traditional MANOVA when the individuals selected in the study is heterogeneous. The repeated measures MANOVA should be used when several dependent variables (DVs) measure different aspects of some cohesive theme. Such cohesive concept may be personality (Extraversion, Psychoticism, and Neuroticism), health (blood pressure, heart rate, and vital capacity), product features (economy, comfort, and attractiveness), fitness (cardio respiratory endurance, flexibility, and strength), nature (extrovert, optimism, and creativity), academic achievement (English, Maths, and Commerce) and so on.

One of the important considerations in using the repeated measures MANOVA is to choose the DVs carefully in the study. Highly correlated DVs weaken the power of the analysis because in that case it does not make sense to use the redundant information. On the other hand, if the DVs are not correlated, MANOVA has nothing to offer. The DVs should have moderate correlation among themselves. As a thumb rule, the correlations among the DVs should be in the range of 0.3 to 0.7.

Word of Caution: Often researcher is tempted to use repeated measures MANOVA even if dependent variables do not explain a cohesive theme. For instance, if the set of DVs include blood pressure, extraversion, and flexibility, there is no point of using the MANOVA technique even if the correlations among them are in between the range 0.3 to 0.7 because in that situation selection of DVs has no theoretical justification and the interpretation would be out of place. In such situations consider using a series of univariate ANOVAs (one for each DV).

WHY TO USE REPEATED MEASURES MANOVA?

The choice of repeated measures MANOVA depends upon the following considerations:

1. Using repeated measures MANOVA depends upon the research question being investigated. By taking many dependent variables the chances for finding group difference increases. In many situations such a test is more powerful than the univariate ANOVA, but this need not necessarily be true always.

2. In many situations dependent variables belonging to the same domain are highly correlated with each other, thus in those situations findings from separate repeated ANOVAs will be redundant and difficult to integrate.

3. Repeated measures MANOVA takes into account the intercorrelations among DVs.

4. Another consideration in using repeated measures MANOVA is that, none of the individual ANOVAs may produce a significant effect on the DV, but if combined they might. This suggests that the variables are more meaningful if taken together than considered separately.

5. By using MANOVA, the family-wise error rate (α) can be controlled in the experiment. Instead, conducting multiple ANOVAs increases the chance for family-wise error rate (α) and increases the odds of finding significant value of F due to chance with the repeated use of the same sample of data.

6. The sphericity assumption in repeated measures ANOVA is often violated, whereas the repeated measure MANOVA does not require this assumption.

ASSUMPTIONS

In using one-way repeated measures MANOVA certain assumptions need to be satisfied. You get valid conclusions only when these assumptions hold true. If some of these assumptions fail, there is always a solution to overcome this. Most of these assumptions can be checked by using the SPSS software, whereas others are design issue. The assumptions for this design are as follows:

a. *Data type*. There should be two or more continuous dependent variables and one categorical independent variable.

b. *Sample Size*. The number of observations must be higher than the number of dependent variables. It is recommended to have minimum of 20 sample size in this design.

c. *Independence of Measurement*. The measurement obtained on each subject must be independent to each other.

d. *Missing Data*. This design requires complete data for all the subjects. In other words, no missing observation for any subject should be there in the experiment.

e. *Outliers*. No outlier should exist in each group of the independent variable for any of the dependent variables. This can be checked by using the Box Plot in the SPSS.

f. *Linearity*. All DVs are linearly related among themselves in each group of the independent variable. This assumption can be tested by means of scatter plots and bi-variate correlations using SPSS.

g. *Normality*. There should be multivariate normality. To test this assumption one may test the normality of each dependent variable for each of the levels of the independent variable by using the Shapiro–Wilk test in SPSS. In other words, if the independent variable has three levels and there are three dependent variables in the design, then all the nine set of data in the design must follow the normal distribution. If normality is violated MANOVA procedures are robust enough to type I error but power of the test is sacrificed.

h. *Multicollinearity*. There should be no multicollinearity among the DVs. If correlations among DVs are more than 0.9, multicollinearity exists and in that case one of the variables need to be dropped from the analysis.

i. *Sphericity*. There should be no sphericity in data. In other words, difference scores computed between two levels of a within-subjects factor must have the same variance for the comparison of any two levels. This assumption is used while solving univariate ANOVA as a follow-up test in MANOVA. The sphericity assumption is tested only in a situation where the independent factor has more than two levels. This assumption is tested by means of Mauchly's W test in SPSS.

APPLICATION

One way repeated measures MANOVA design can be used to investigate varieties of research issues in different disciplines. Some of the specific applications are listed below:

1. A psychologist may wish to investigate the trend in personality transformation during one year of training classes in communication skill. Three personality traits, Extraversion, Psychoticism, and Neuroticism of the subjects, may be tested at regular intervals during training program. In this case the time would be an independent variable (within-subjects) with testing points as its levels whereas personality parameters; Extraversion, Psychoticism, and Neuroticism would be dependent variables.

2. A researcher may wish to investigate as to which naturopathy intervention is more effective in improving mood state. The subjects may be exposed to pranayama, meditation, and relaxation exercise. The subjects may be tested on three dimensions of the mood state; confusion, depression, and fatigue, after each intervention.

3. An educational consultant may wish to investigate performance trend of subjects during a training programme for a competitive examination. The performance of the subjects on numerical aptitude, reasoning, and English comprehension may be tested after equal interval of time during the training programme.

LAYOUT DESIGN

Layout of one-way repeated measures MANOVA design depends upon whether the levels of independent factor are different treatments or time periods. Thus, the two different kinds of layouts are as follows:

Case I: When Levels of Within-Subjects Factor are Different Treatments

In this layout design, each subject is tested on multiple dependent variables in each treatment. Sufficient time gap is given between testing under any two treatments in order to done away the learning or fatigue effect which may result due to earlier treatment. Further, counterbalancing is done to remove order effect. Let us consider an example of mood state discussed above in which six randomly selected subjects are exposed to three different naturopathy treatments; pranayama, meditation, and relaxation exercise. Each treatment is given for a week. These subjects are tested on three dimensions of the mood state; confusion, depression, and fatigue, after each intervention. Here naturopathy is a within-subjects factor and the three dependent variables are confusion, depression, and fatigue. As per testing protocol, in the first phase the randomly selected subjects, say, S2 and S5 are given pranayama, the subjects S1 and S6 are given meditation, and the subjects S3 and S4 are exposed to relaxation exercise for a week. Subjects are tested for their mood state parameters, confusion, depression, and fatigue, after each treatment. Similarly random assignment of treatments to the subjects is done in the second and third phase of the treatment. Sufficient gap between each phase of treatment is kept in order to avoid the learning or fatigue effect. Different kinds of layout design are possible by randomizing treatments to the subjects in sample. While doing the randomization the researcher must ensure that all subjects receive each treatment. The layout of this design is shown in Figure 7.1.

Testing protocol
Treatment: Naturopathy intervention

	Pranayama			Meditation			Relaxation Excercise		
	Confusion	Depression	Fatigue	Confusion	Depression	Fatigue	Confusion	Depression	Fatigue
First phase testing	S2	S2	S2	S1	S1	S1	S3	S3	S3
	S5	S5	S5	S6	S6	S6	S4	S4	S4
Second phase testing	S1	S1	S1	S3	S3	S3	S2	S2	S2
	S6	S6	S6	S4	S4	S4	S5	S5	S5
Third phase testing	S3	S3	S3	S2	S2	S2	S1	S1	S1
	S4	S4	S4	S5	S5	S5	S6	S6	S6

Figure 7.1 Layout of the one-way repeated measures MANOVA design having three levels of independent factor as different treatments

Testing protocol

Treatment: Time

Initial			2 week			4 week		
Numerical Aptitude	Reasoning	English comprehension	Numerical Aptitude	Reasoning	English comprehension	Numerical Aptitude	Reasoning	English comprehension
S1	S1	S1	S1	S1	S1	S1	S1	S1
S2	S2	S2	S2	S2	S2	S2	S2	S2
S3	S3	S3	S3	S3	S3	S3	S3	S3
S4	S4	S4	S4	S4	S4	S4	S4	S4
S5	S5	S5	S5	S5	S5	S5	S5	S5
S6	S6	S6	S6	S6	S6	S6	S6	S6

Figure 7.2 Layout of the one-way repeated measures MANOVA design having three treatment levels as time durations

Case II: When Levels of Within-Subjects Factor are Different Time Durations

In this layout design, all subjects are repeatedly tested on all the dependent variables at different time duration during treatment. The focus of the researcher here is to see the trend of improvement in different duration. Consider the above-mentioned example of investigating performance trend by the educational consultant. Let us consider that the six randomly selected subjects are given a training programme for appearing in a competitive examination. These subjects are tested for their numerical aptitude, reasoning, and English comprehension before training and after 2 weeks and 4 weeks of the training programme. The layout of this repeated measure MANOVA design can be shown by the Figure 7.2.

STEPS IN SOLVING ONE-WAY REPEATED MEASURES MANOVA

In this design a researcher has two objectives; firstly, to examine the effect of independent factor on a group of dependent variables, and secondly, to identify the differences between individual groups for each of the dependent measure with pair-wise comparison of means. Following steps are used in solving this design:

1. Check the assumption of data type, independence of measurements, outliers, linearity, normality, and multicollinearity.
2. Describe layout of the design
3. Specify research questions to be investigated
4. Formulate hypotheses to be tested
5. Decide family-wise error rate (α) for the study
6. Use SPSS commands to generate the following outputs:
   a. Descriptive Statistics

ILLUSTRATION 167

b. MANOVA table containing Wilks' Lambda and other multivariate test for investigating within-subjects effects on a group of dependent variables.
7. If Wilks' statistic is significant interpret the following outputs generated in SPSS to test the within-subjects effects on each dependent variable:
   a. Mauchly's test of sphericity for each dependent variable.
   b. One-way repeated measures ANOVA for testing the effect of independent variable on each dependent variable separately.
   c. Pair-wise comparisons of means for each dependent variable.
   d. Means plot for each dependent variable.
8. Report the findings.

The procedure involved in solving this design shall be explained by means of the below-mentioned illustration.

ILLUSTRATION

A study was conducted to see the effect of time of the day on the student's performance in different subjects. Ten school students were randomly selected from the population of interest and were given the test in Mathematics, English, and Reasoning. The marks so obtained by the students are shown in the Table 7.1. Let's see how one-way repeated measures MANOVA can be applied to discuss the findings in this analysis.

Layout Design

In this illustration effect of time of testing is to be seen on the combined effect of the performance on different subjects; Maths, English, and Reasoning. All ten subjects

Table 7.1 Marks Obtained by the Students in Different Subjects at Different Times of the Day

Time of the day								
Morning (7 AM)			Afternoon (1 PM)			Evening (7 PM)		
Maths	English	Reasoning	Maths	English	Reasoning	Maths	English	Reasoning
12	12	15	15	15	11	17	14	12
13	14	16	17	13	12	16	12	10
14	10	17	18	14	14	15	15	15
13	9	15	15	14	13	16	16	12
14	8	17	14	13	11	14	13	14
15	11	15	18	12	10	16	15	15
13	10	14	17	15	9	15	13	10
12	13	15	15	12	8	13	12	13
13	12	13	16	15	11	15	16	12
15	11	14	18	16	12	16	15	13

Testing protocol

Treatment: Time of the day

	Morning			Afternoon			Evening		
	Maths	English	Reasoning	Maths	English	Reasoning	Maths	English	Reasoning
First phase testing	S1 S3	S1 S3	S1 S3	S2 S4	S2 S4	S2 S4	S5 S6	S5 S6	S5 S6
Second phase testing	S2 S4	S2 S4	S2 S4	S5 S6	S5 S6	S5 S6	S1 S3	S1 S3	S1 S3
Third phase testing	S5 S6	S5 S6	S5 S6	S1 S3	S1 S3	S1 S3	S2 S4	S2 S4	S2 S4

Figure 7.3   Layout of the one-way repeated measures MANOVA design in the illustration

are tested in all the three treatment conditions (time of the day). The design used in this illustration is a one-way repeated measures MANOVA, the layout of which can be shown by the Figure 7.3.

In this design all subjects are tested on all the three dependent variables (Maths, English, and Reasoning), but not in a particular sequence. It can be seen from the testing protocol in Figure 7.3 that in the first phase of testing two subjects, S1 and S3 are tested for their performance in the morning, S2 and S4 in the afternoon, and S5 and S6 in the evening. This order has been randomized in the second and third phase of testing. Readers should note that the sufficient time gap should be given between any two phases of testing in order to done away the effect of learning or fatigue, whereas random allocation of treatment to the subjects should be done for counterbalancing. One can randomize the allocation of treatment conditions to the subjects in a different manner also; the only condition is that each of the ten subjects should be tested in all three treatment conditions.

Research Questions

The following three research questions shall be investigated in this illustration.

a. Whether time of testing affects the student's academic performance in general?
b. Whether the time of testing affects the student's performance in each of the subject; Maths, English, and Reasoning?
c. Which time of the day improves performance of the student's in each subject?

Hypotheses Construction

a. To investigate the first research question following hypotheses shall be tested:

H_0 : There is no difference between group mean vectors of

the student's performance in three different time of testing.

ILLUSTRATION 169

against

H_1 : At least one group mean vector differs.

The above-mentioned null hypothesis can be written mathematically as follows:

$$H_0 : \begin{bmatrix} \mu_{\text{Maths}} \\ \mu_{\text{English}} \\ \mu_{\text{Reasoning}} \end{bmatrix}_{\text{Morning}} = \begin{bmatrix} \mu_{\text{Maths}} \\ \mu_{\text{English}} \\ \mu_{\text{Reasoning}} \end{bmatrix}_{\text{Afternoon}} = \begin{bmatrix} \mu_{\text{Maths}} \\ \mu_{\text{English}} \\ \mu_{\text{Reasoning}} \end{bmatrix}_{\text{Evening}} \qquad (7.1)$$

b. To investigate the second research question the following hypotheses shall be tested

H_0 : There is no difference between group means of student's

performance in a subject in three different time of testing.

against

H_1 : At least one group mean differs.

These hypotheses can be written mathematically as shown below:

$$H_0 : \mu_{\text{Morning}} = \mu_{\text{Afternoon}} = \mu_{\text{Evening}}$$

against

H_1 : At least any one group mean differs $\qquad (7.2)$

Testing the first set of hypotheses shall answer the first research question as to whether the combined performance of students differs in three different time of testing. This is done by using repeated MANOVA technique in which Wilks' test or Pillai test is used more often to test the null hypothesis. Generally, Wilks' test is used in MANOVA analysis, but if the sample size decreases or the cell sizes are unequal then Pillai's test is better option.

Wilks' lambda gives the proportion of variance in the combination of dependent variables (Maths, English, and Reasoning) that is unaccounted for by the independent variable (the time of testing). Thus, lesser the value of Wilks' lambda more is the variation in the group mean vectors. Wilks' lambda statistic follows approximately F distribution, hence F statistic is also computed in the output of MANOVA. Thus, for the maximum group difference Wilks' lambda should be low and the corresponding F value should be significant.

If the F value associated with Wilks' lambda is significant, then the second set of hypothesis mentioned above is tested for each dependent variable using univariate ANOVA. If F value in any univariate ANOVA is significant, then the pair-wise comparison of means shall be done by applying the paired t-test using Bonferroni correction. This pair-wise comparison of means shall address the third research question.

Level of Significance

In this illustration, family-wise error rate (α) shall be taken as 0.05. If the Wilk test is significant in MANOVA then as a follow-up test three univariate ANOVA shall be done in order to see the effect of time of testing on each dependent variable. This will inflate the family-wise error rate. To compensate this, the F value in the univariate ANOVA shall be tested at the significance level 0.017 (= 0.05/3).

Solving One-Way Repeated Measures MANOVA Design with SPSS

The detail procedure involved in one-way repeated measures MANOVA shall be discussed by solving the problem discussed in the illustration using SPSS. Readers should carefully note the format of data file in this design. In **Variable View** define all the nine variables as shown below:

Maths_M
Maths_A
Maths_E
English_M
English_A
English_E
Reasoning_M
Reasoning_A
Reasoning_E

In **Data View** enter the data of the Table 7.1 column-wise as shown in the Figure 7.4. After preparing the data file, click on the following sequence of commands as shown in Figure 7.5 to get the screen shown in Figure 7.6 for defining independent and dependent variables.

Analyze ⟶ General linear model ⟶ Repeated measures

Factor 1 is written in the "Within-Subject Factor Name" section by default as shown in Figure 7.6(a). Replace this by the independent variable *Time* (within-subjects) and write the number of levels as 3 as there are three levels (Morning, Afternoon, and

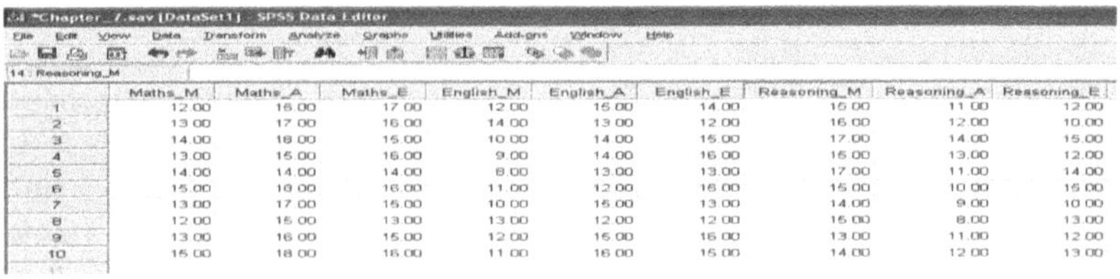

	Maths_M	Maths_A	Maths_E	English_M	English_A	English_E	Reasoning_M	Reasoning_A	Reasoning_E
1	12.00	16.00	17.00	12.00	15.00	14.00	15.00	11.00	12.00
2	13.00	17.00	16.00	14.00	13.00	12.00	16.00	12.00	10.00
3	14.00	18.00	15.00	10.00	14.00	15.00	17.00	14.00	15.00
4	13.00	15.00	16.00	9.00	14.00	16.00	15.00	13.00	12.00
5	14.00	14.00	14.00	8.00	13.00	13.00	17.00	11.00	14.00
6	15.00	18.00	16.00	11.00	12.00	16.00	15.00	10.00	15.00
7	13.00	17.00	15.00	10.00	15.00	13.00	14.00	9.00	10.00
8	12.00	15.00	13.00	13.00	12.00	12.00	15.00	8.00	13.00
9	13.00	16.00	15.00	12.00	15.00	16.00	13.00	11.00	12.00
10	15.00	18.00	15.00	11.00	16.00	15.00	14.00	12.00	13.00

Figure 7.4 Data format in one-way repeated measure MANOVA

ILLUSTRATION   171

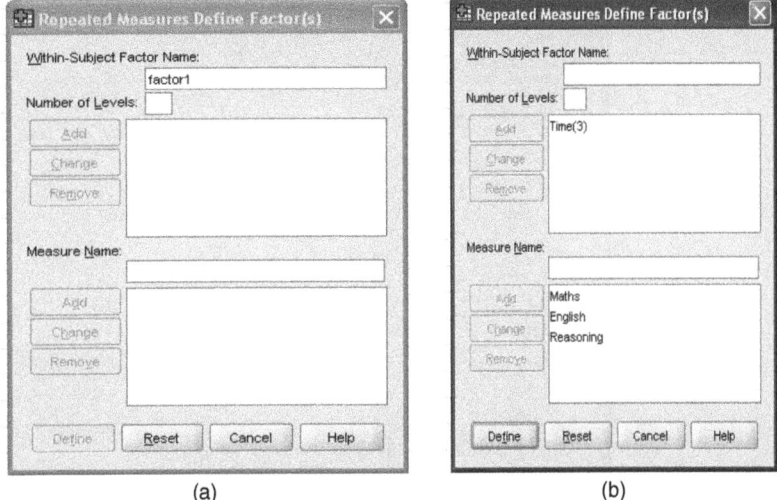

Figure 7.5 Screen for initiating command for one-way repeated measure MANOVA

Figure 7.6 Screen showing options for defining independent and dependent variables

Evening). Click on **Add** to enter this information into the box. Write the name of all the three dependent variables (*Maths, English* and *Reasoning*) one by one in "Measure Name" section by clicking on **Add.** After defining variables the screen shall look like as shown in Figure 7.6(b).

Clicking on **Define** command shall take you to the screen as shown in Figure 7.7(a) for selecting variables defined in data file. Transfer all nine variables from the left panel to the "Within-Subjects Variables" section of the screen in the right panel by using the arrow key. After selecting variables the screen shall look like as shown in Figure 7.7(b).

After selecting the variables, option needs to be defined for generating means plot in univariate-repeated ANOVA for each dependent variable. Click on the **Plots** command to get the screen as shown in Figure 7.8. Shift *Time* variable from the "Factors"

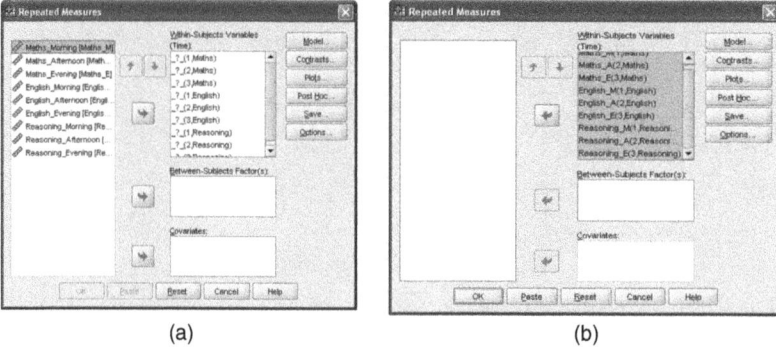

(a)	(b)

Figure 7.7   Screen showing option for selecting variables defining all treatment combinations

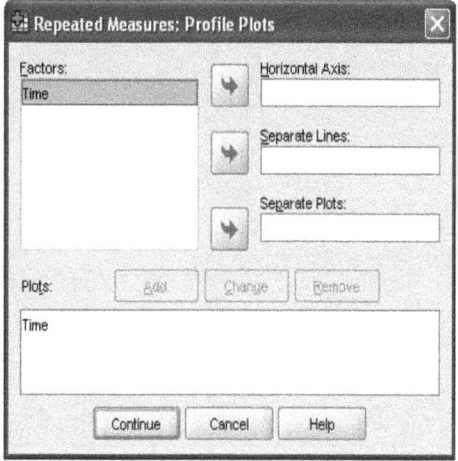

Figure 7.8   Screen showing option for means plot

section to the "Horizontal Axis" section for generating means plot of the independent variable Time (within subjects) for each dependent variable separately.

These plots are used to compare the effect of different levels of Time (within-subjects) on each of the dependent variable separately. Click on **Add** and **Continue.** This will take you back to the screen as shown in Figure 7.7(b). Click on **Options** command to get the screen as shown in Figure 7.9 for selecting options to generate all required outputs in the analysis.

Shift *Time* variable from "Factor(s) and Factor Interactions" section to the "Display Means for" section. This will generate marginal means of Time in each dependent variable.

Check the option 'Compare Mean effects'. Select 'Bonferroni' correction by clicking on the sign ∇ mentioned on the "LSD(none)" tag. This will generate the outputs for testing the effect of time on each dependent variable separately. Since there is no post-hoc test available for within-subjects factor, Bonferroni correction has to be used

ILLUSTRATION   173

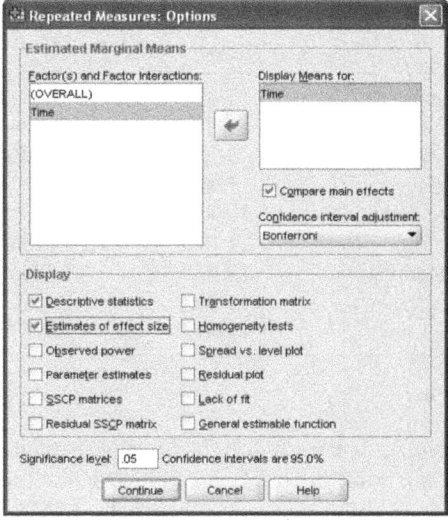

Figure 7.9   Screen showing option for generating various outputs in one-way repeated measures MANOVA

to compensate the inflated family-wise error rate (α) due to multiple comparisons. Readers are advised to refer to the Chapter 3 for detail discussion on it.

Check 'Descriptive statistics' option for computing mean and standard deviation of the data in each treatment condition. Ensure that the level of significance is selected as 0.05. Check 'Estimates of effect size' option. Let all other options remain as it is. Click on **Continue** to get back to the screen shown in Figure 7.7(b). Click on **OK** to get the output.

SPSS Output and Interpretation

The SPSS generates plenty of outputs in one-way repeated measure MANOVA, but only following relevant outputs need to be picked up for interpretation.

1. Descriptive statistics (Table 7.2)
2. Multivariate tests of within-subjects effects (Table 7.5)
3. Mauchly's test of sphericity (Table 7.6)
4. Univariate tests: one-way repeated measures ANOVA for each dependent variable (Table 7.7)
5. Estimates of marginal means in different groups (Table 7.8)
6. Pair-wise comparison of marginal means in each dependent variable (Table 7.9)
7. Marginal means plot of Maths (Figure 7.13)
8. Marginal means plot of English (Figure 7.14)
9. Marginal means plot of Reasoning (Figure 7.15)

Table 7.2   Descriptive Statistics

	Mean	SD	N
Maths_Morning	13.4000	1.07497	10
Maths_Afternoon	16.3000	1.49443	10
Maths_Evening	15.3000	1.15950	10
English_Morning	11.0000	1.82574	10
English_Afternoon	13.9000	1.37032	10
English_Evening	14.1000	1.52388	10
Reasoning_Morning	15.1000	1.28668	10
Reasoning_Afternoon	11.1000	1.79196	10
Reasoning_Evening	12.6000	1.77639	10

The output in Table 7.5 is used to test the hypothesis related to the first research question. It tests whether group difference is significant or not across all the dependent variables. If no significant difference is observed in this MANOVA table, the analysis stops and no further analysis is done. However, if the group difference is significant, then one needs to investigate univariate repeated measures ANOVA for each dependent variable separately. Have some patience; the discussion in the following sections shall make the concept clear.

Descriptive Statistics   The mean and SD for the observations on all the dependent variables in each group have been generated in the Table 7.2. These statistics can be used to compare the mean and standard deviation in each cell. The means of different groups are used to prepare means plots for comparing the effect of different levels of time on each dependent variable separately.

Testing Assumptions

Before using one-way repeated measures MANOVA, it is essential to test its assumptions. Following assumptions shall be tested by using the data file as shown in Figure 7.4 in SPSS.

Testing Correlation   One of the key assumptions of this analysis is that there should be reasonable relationship between the dependent variables in each level of the independent variable. The correlation matrix in Table 7.3 has been obtained by clicking on the following sequence of commands:

Analyze ⟶ Correlate ⟶ Bivariate

The MANOVA test will not be appropriate if the correlations among them is very low and in that case univariate analysis should be done to investigate the effect of independent variable on each of the dependent variable separately. However, if the correlation among dependent variables is very high (>0.9), it is a sign of multicollinearity and in that case one of the variables should be dropped. Let us examine the relationship here. Table 7.3 shows that the correlation between Maths_M

Table 7.3 Correlations

		Maths_M	Maths_A	Maths_E	English_M	English_A	English_E	Reason_M	Reason_A	Reason_E
Maths_M	Pearson Correlation	1	0.609	0.160	**−0.396**	0.030	0.380	**0.129**	0.323	0.559
	Sig. (2-tailed)		0.062	0.658	0.257	0.934	0.279	0.723	0.363	0.093
	N	10	10	10	10	10	10	10	10	10
Maths_A	Pearson correlation	0.609	1	0.327	0.204	**0.179**	0.229	−0.133	**0.236**	0.134
	Sig. (2-tailed)	0.062		0.356	0.573	0.621	0.524	0.714	0.511	0.712
	N	10	10	10	10	10	10	10	10	10
Maths_E	Pearson correlation	0.160	0.327	1	0.105	0.441	0.421	−0.171	0.465	**−0.205**
	Sig. (2-tailed)	0.658	0.356		0.773	0.203	0.225	0.636	0.175	0.570
	N	10	10	10	10	10	10	10	10	10
English_M	Pearson correlation	−0.396	0.204	0.105	1	−0.133	−0.359	−0.284	−0.272	−0.377
	Sig. (2-tailed)	0.257	0.573	0.773		0.714	0.308	0.427	0.448	0.283
	N	10	10	10	10	10	10	10	10	10
English_A	Pearson correlation	0.030	0.179	0.441	−0.133	1	0.431	−0.498	0.321	−0.338
	Sig. (2-tailed)	0.934	0.621	0.203	0.714		0.214	0.143	0.365	0.340
	N	10	10	10	10	10	10	10	10	10
English_E	Pearson correlation	0.380	0.229	0.421	−0.359	0.431	1	−0.346	0.525	0.345
	Sig. (2-tailed)	0.279	0.524	0.225	0.308	0.214		0.328	0.119	0.329
	N	10	10	10	10	10	10	10	10	10
Reason_M	Pearson correlation	0.129	−0.133	−0.171	−0.284	−0.498	−0.346	1	0.381	0.408
	Sig. (2-tailed)	0.723	0.714	0.636	0.427	0.143	0.328		0.278	0.241
	N	10	10	10	10	10	10	10	10	10
Reason_A	Pearson correlation	0.323	0.236	0.465	−0.272	0.321	0.525	0.381	1	0.188
	Sig. (2-tailed)	0.363	0.511	0.175	0.448	0.365	0.119	0.278		0.602
	N	10	10	10	10	10	10	10	10	10
Reason_E	Pearson correlation	0.559	0.134	−0.205	−0.377	−0.338	0.345	0.408	0.188	1
	Sig. (2-tailed)	0.093	0.712	0.570	0.283	0.340	0.329	0.241	0.602	
	N	10	10	10	10	10	10	10	10	10

Bold face indicates that the effect is significant at 5% level.

and English_M is −0.396, and that of between Maths_M and Reason_M is 0.129. In afternoon group correlation between Maths and English and Maths and Reasoning are 0.179 and 0.236, respectively. Similarly in the evening group correlation between Maths and English and Maths and Reasoning are 0.421 and −0.205, respectively. None of these correlations is significant, hence MANOVA analysis may not result in any logical conclusion. This situation has arisen in this illustration because the number of sample is only six. If twenty samples would have been taken as recommended for this design, then the correlation might have been significant and in that case it would have been logical to club the three dependent variables (Maths, English and Reasoning) provided theory supports clubbing these three variables together in assessing the academic performance of the students. Thus, in such situations the researchers must reinvestigate the selection of dependent variables in their study for clubbing together to draw any cohesive conclusion.

Testing Normality The normality assumption shall be tested by using the Shapiro–Wilks' test in SPSS. By using the below mentioned sequence of commands and selecting the option 'Normality plots with test' after clicking on the **Options** command, the output shown in Table 7.4 has been obtained.

Analyze ⟶ Descriptive statistics ⟶ Explore

For details readers are advised to refer to Chapter 3. Since none of the value of Shapiro–Wilk statistic is significant ($p > 0.05$), the assumption of normality holds true for all the nine data sets.

Testing Outliers For using this design no outlier should be present among the data in each cell. The outlier can be detected in the data by using the Box plot. The box plots for each data set shown in Figures 7.10–7.12 have been obtained by clicking on the following commands in sequence and checking the option 'Outliers' after clicking on the **Statistics** command.

Analyze ⟶ Descriptive statistics ⟶ Explore

The box plot shown in Figures 7.10(c) reveals that only 8th score of mathematics performance obtained in the evening seems to be outlier. It is up to the researcher to

Table 7.4   Test of Normality

Treatment Groups	Shapiro–Wilk Statistic	df	Sig.
Maths_M	0.892	10	0.177
Maths_A	0.882	10	0.138
Maths_E	0.916	10	0.328
English_M	0.984	10	0.982
English_A	0.926	10	0.410
English_E	0.895	10	0.193
Reasoning_M	0.924	10	0.392
Reasoning_A	0.976	10	0.937
Reasoning_E	0.918	10	0.338

ILLUSTRATION 177

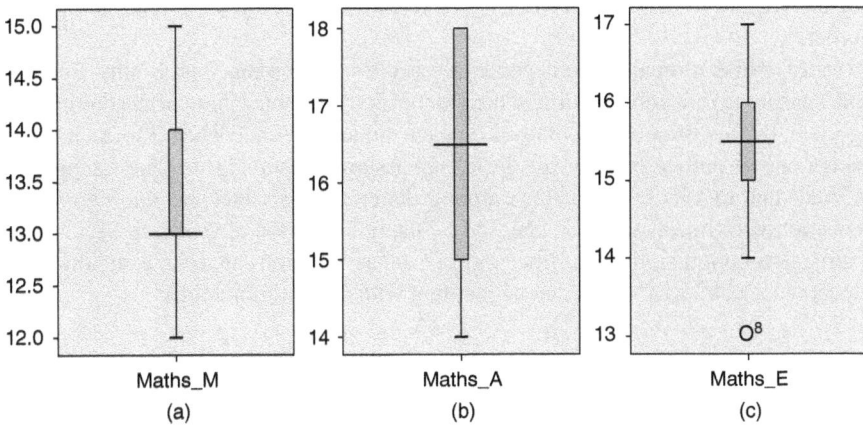

Figure 7.10 Box plots of Maths scores

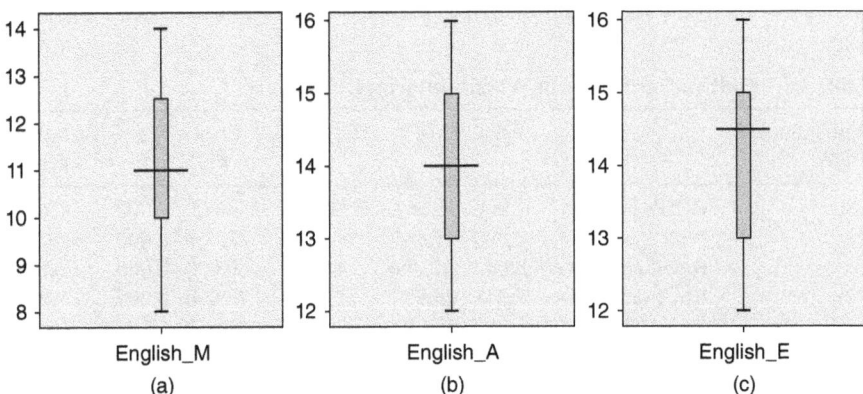

Figure 7.11 Box plots of English scores

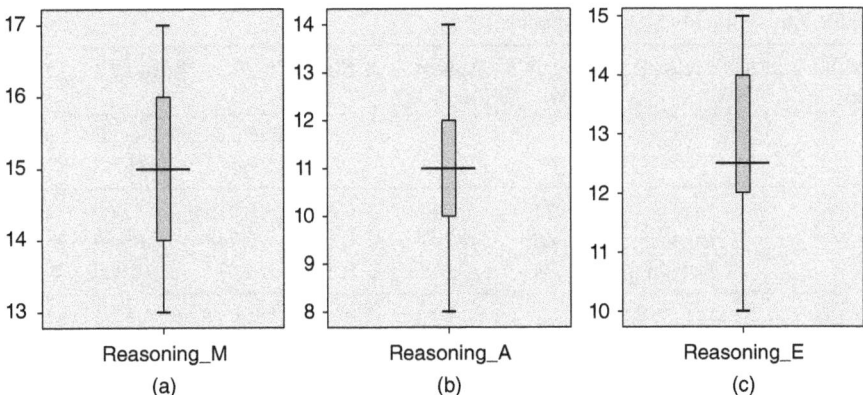

Figure 7.12 Box plots of Reasoning scores

decide whether to keep this data or not. Except this score no other data set has any outlier.

In this illustration all three dependent variables (performance on Maths, English, and Reasoning) are continuous and the independent variable (Time of the day) is categorical; the measurements obtained on each subject are independent to each other, except one no outlier exists hence the design assumptions hold true. The sample size is small due to which correlations among dependent variables are not significant. Assumption of linearity can be checked by using the option of Graph in SPSS. Normality assumption holds true. Since most of the assumptions are true, using one-way repeated measure MANOVA can be justified with certain limitations.

Multivariate Testing

Table 7.5 is the output for testing the null hypothesis (7.1) related to group difference across all the three dependent variables. Since the value of Wilks' Lambda is 0.099 with F value 11.585 ($p < 0.001$), the null hypothesis is rejected. It is thus concluded that there is a significant multivariate effect across within-subjects time point

Table 7.5   Multivariate[a,b] Tests of Within-Subjects Effects

Within Subjects Effect		Value	F	Hypothesis df	Error df	Sig.	Partial Eta Squared
Time	Pillai's trace	1.000	5.662	6.000	34.000	0.000	0.500
	Wilks' lambda	0.099	11.585[c]	6.000	32.000	0.000	0.685
	Hotelling's trace	8.066	20.166	6.000	30.000	0.000	0.801
	Roy's largest root	7.941	44.999[d]	3.000	17.000	0.000	0.888

[a]Design: Intercept
Within Subjects Design: Time
[b]Tests are based on averaged variables.
[c]Exact statistic
[d]The statistic is an upper bound on F that yields a lower bound on the significance level.

Table 7.6   Mauchly's Test of Sphericity[a]

Within Subjects Effect	Measure	Mauchly's W	Approx. Chi-Square	df	Sig.	Epsilon[b] Greenhouse–Geisser	Huynh–Feldt	Lower-Bound
Time	Maths	0.911	0.743	2	0.690	0.918	1.000	0.500
	English	0.662	3.297	2	0.192	0.748	0.863	0.500
	Reasoning	0.854	1.262	2	0.532	0.873	1.000	0.500

Test the null hypothesis that the error covariance matrix of the orthonormalized transformed-dependent variables is proportional to an identity matrix.
[a]Design: Intercept
Within Subjects Design: Time
[b]May be used to adjust the degrees of freedom for the averaged tests of significance. Corrected tests are displayed in the Tests of Within-Subjects Effects table.

Table 7.7 Univariate Tests: One-Way Repeated Measure ANOVA for Each Dependent Variable

Source	Measure		Type III SS	df	Mean Square	F	Sig. (p-Value)	Partial Eta Squared
Time	Maths	Sphericity assumed	43.400	2	21.700	21.781	**0.000**	0.708
		Greenhouse–Geisser	43.400	1.837	23.626	21.781	0.000	0.708
		Huynh–Feldt	43.400	2.000	21.700	21.781	0.000	0.708
		Lower-bound	43.400	1.000	43.400	21.781	0.001	0.708
	English	Sphericity assumed	60.200	2	30.100	11.335	**0.001**	0.557
		Greenhouse–Geisser	60.200	1.495	40.267	11.335	0.002	0.557
		Huynh–Feldt	60.200	1.726	34.888	11.335	0.001	0.557
		Lower-bound	60.200	1.000	60.200	11.335	0.008	0.557
	Reasoning	Sphericity assumed	81.667	2	40.833	21.832	**0.000**	0.708
		Greenhouse–Geisser	81.667	1.745	46.794	21.832	0.000	0.708
		Huynh–Feldt	81.667	2.000	40.833	21.832	0.000	0.708
		Lower-bound	81.667	1.000	81.667	21.832	0.001	0.708
Error (Time)	Maths	Sphericity assumed	17.933	18	0.996			
		Greenhouse–Geisser	17.933	16.533	1.085			
		Huynh–Feldt	17.933	18.000	0.996			
		Lower-bound	17.933	9.000	1.993			
	English	Sphericity assumed	47.800	18	2.656			
		Greenhouse–Geisser	47.800	13.455	3.553			
		Huynh–Feldt	47.800	15.530	3.078			
		Lower-bound	47.800	9.000	5.311			
	Reasoning	Sphericity assumed	33.667	18	1.870			
		Greenhouse–Geisser	33.667	15.707	2.143			
		Huynh–Feldt	33.667	18.000	1.870			
		Lower-bound	33.667	9.000	3.741			

Bold face indicates that the effect is significant at 5% level.

Table 7.8 Estimates of Marginal Means in Different Groups

Measure	Time	Mean	Std. Error	95% Confidence Interval	
				Lower Bound	Upper Bound
Maths	Morning	13.400	0.340	12.631	14.169
	Afternoon	16.300	0.473	15.231	17.369
	Evening	15.300	0.367	14.471	16.129
English	Morning	11.000	0.577	9.694	12.306
	Afternoon	13.900	0.433	12.920	14.880
	Evening	14.100	0.482	13.010	15.190
Reasoning	Morning	15.100	0.407	14.180	16.020
	Afternoon	11.100	0.567	9.818	12.382
	Evening	12.600	0.562	11.329	13.871

on student's performance (Maths, English, and Reasoning). Also partial eta square for Time is 0.685 which is considered to be quite high, hence the effect of Time of testing in assessing the performance is of immense practical value. In order to know how the time of testing affects individual dependent variable, independent repeated measures ANOVA results shall be examined.

Table 7.9 Pair-Wise Comparison of Marginal Means in Each Dependent Variable

Measure	Time (I)	Time (J)	($I - J$)	Std. Error	Sig[a]	Lower Bound	Upper Bound
Maths	Morning	Afternoon	−2.900[*]	0.379	**0.000**	−4.011	−1.789
		Evening	−1.900[*]	0.458	**0.007**	−3.244	−0.556
	Afternoon	Morning	2.900[*]	0.379	**0.000**	1.789	4.011
		Evening	1.000	0.494	0.221	−0.450	2.450
	Evening	Morning	1.900[*]	0.458	**0.007**	0.556	3.244
		Afternoon	−1.000	0.494	0.221	−2.450	0.450
English	Morning	Afternoon	−2.900[*]	0.767	**0.013**	−5.149	−0.651
		Evening	−3.100	0.875	0.019	−5.667	−0.533
	Afternoon	Morning	2.900[*]	0.767	**0.013**	0.651	5.149
		Evening	−0.200	0.490	1.000	−1.637	1.237
	Evening	Morning	3.100	0.875	0.019	0.533	5.667
		Afternoon	0.200	0.490	1.000	−1.237	1.637
Reasoning	Morning	Afternoon	4.000[*]	0.558	**0.000**	2.364	5.636
		Evening	2.500[*]	0.543	**0.004**	0.908	4.092
	Afternoon	Morning	−4.000[*]	0.558	**0.000**	−5.636	−2.364
		Evening	−1.500	0.719	0.200	−3.608	0.608
	Evening	Morning	−2.500[*]	0.543	**0.004**	−4.092	−0.908
		Afternoon	1.500	0.719	0.200	−0.608	3.608

Based on estimated marginal means
[*]The mean difference is significant at the 0.017 level.
[a]Adjustment for multiple comparisons: Bonferroni.
Bold face indicates that the effect is significant at 5% level.

ILLUSTRATION 181

Univariate Testing

Before analyzing univariate results to compare the performance of the subjects in three time groups for each dependent variable, it is essential to test the sphericity first because time is a within-subjects factor.

Testing Sphericity Table 7.6 shows that Mauchly's statistic is not significant for any of the three dependent variables as p values associated with chi-square for all the dependent variables are more than 0.05. It may therefore be concluded that the sphericity assumption is not violated in any of the dependent variable, hence no correction is required for sphericity.

Since sphericity has not been violated, the results shown in front of the 'Sphericity Assumed' row in Table 7.7 shall be discussed to test the hypothesis related to the univariate repeated measures ANOVA. Due to multiple ANOVA the family-wise error rate for testing the significance of F value in univariate ANOVA shall be $0.017 (= 0.05/3)$ instead of 0.05. Table 7.7 shows that the F values (Sphericity Assumed) for Maths, English, and Reasoning are significant because p values associated with these F are less than 0.017. To find as to which time of testing is more productive in each case, pair-wise comparison of marginal means shall be done.

Table 7.8 gives the marginal means and confidence intervals for each cell. This information is useful to have an idea about the variation of the performance in each group and will also be used to interpret the means plot of each dependent variable.

Pair-Wise Comparison of Marginal Means Since post-hoc test cannot be applied for the repeated measures design, no option for comparing the marginal means of different time groups for each of the dependent variable is available for the repeated measures design in SPSS. It uses paired t test for comparing each pair of the marginal group means. Due to multiple comparisons the family-wise error rate inflates, and to compensate this error, Bonferroni correction (for detail see Chapter 2) option has been used in SPSS while comparing (Figure 7.9). Table 7.9 provides the pair-wise comparison of marginal means of different time groups in each dependent variable. Significance of mean difference shall be tested at the significance level 0.017. All significant mean differences have been marked with asterisk (*).

Means Plot of Maths Figure 7.13 shows the means plot of Maths. This means plot has been generated by using the contents of Tables 7.8 and 7.9 for the dependent variable Maths. It can be noted from this figure that the subject's performance is higher during afternoon as well as evening testing in comparison to that of morning testing. However, there is no difference in the Maths performance if tested in the afternoon and evening. It may therefore be concluded that the performance of the students on mathematics is improved if tested either in the afternoon or evening, whereas the performance is least if tested in the morning.

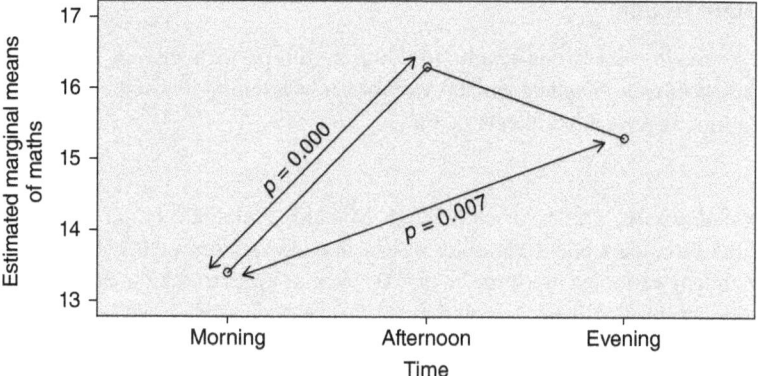

Figure 7.13 Marginal means plot of Maths

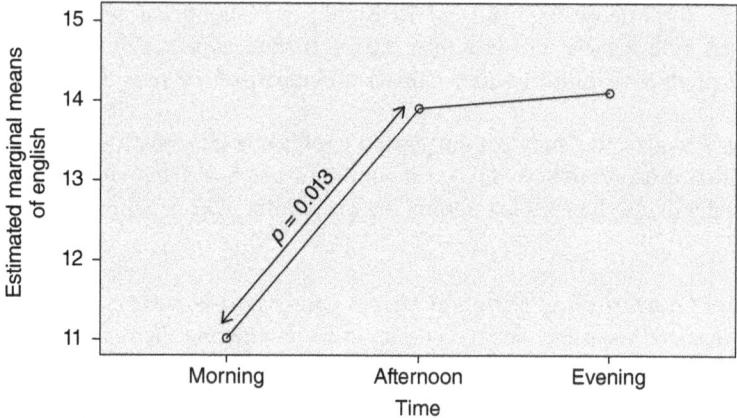

Figure 7.14 Marginal means plot of English

Means Plot of English Figure 7.14 shows the means plot of English. This figure has been generated by using the contents of Tables 7.8 and 7.9 for the dependent variable English. It may be seen in this figure that the subject's performance is higher during afternoon in comparison to that of morning testing. Thus, it may be concluded that the performance of the students on English improves if tested in the afternoon, whereas the performance is least if tested in the morning.

Means Plot of Reasoning Figure 7.15 is a means plot of Reasoning. This figure reveals that the subject's performance in reasoning is the best if tested in the morning in comparison to that of the performance in the afternoon as well as in the evening. There is no difference in the student's performance on reasoning tested in the

ILLUSTRATION   **183**

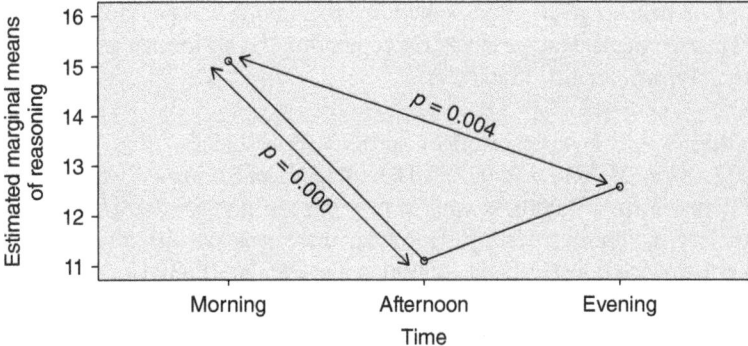

Figure 7.15   Marginal means plot of Reasoning

afternoon and evening. It may therefore be concluded that during the morning time student's performance in reasoning testing is the best.

How to Report the Findings

In this illustration, one-way repeated measure MANOVA was used to investigate the effect of Time (within-subjects) on a group of dependent variables (Maths, English, and Reasoning). The hypotheses were tested at the family-wise error rate of 0.05.

Assumptions   To test the suitability of using one-way repeated measures MANOVA, its assumptions were tested which are as follows:

- Since the independent variable (*Time*) is categorical and dependent variables (marks on Maths, English, and Reasoning) are continuous, assumption about data type holds true.
- Since subject's performance was tested independently with each other, the assumptions of independence also hold true.
- The dependent variables did not seem to be correlated, this assumption was violated (Table 7.3).
- Since Shapiro–Wilk statistic was not significant in each of the nine groups of scores, the assumption of normality is not violated (Table 7.4).
- Except one, no other outlier exists, hence this assumption can also be considered to be true because the outlier is not extreme (Figure 7.10(c)).

Testing Multivariate Effect

- There was a significant multivariate effect across within-subjects time point regardless of subjects (Maths, English, and Reasoning): Wilks' Lambda $= 0.99$, $F(6,32) = 11.585, p < 0.001$.

Testing Univariate Effect   Since sphericity assumption was not violated for the data in all the three dependent variables for comparing the performance across the time, no correction was applied (Table 7.6).

- There was a significant effect across within-subjects time point in Maths, $F(2,18) = 21.781$, $p < 0.017$. The subject's performance was higher during afternoon ($p = 0.000$) as well as evening testing ($p = 0.007$) in comparison to that of morning testing. However, there was no difference in the maths performance if tested in the afternoon and evening (Figure 7.13).
- There was a significant effect across within-subjects time point in English $F(2,18) = 11.335$, $p < 0.017$.

The subject's performance on English was higher during afternoon ($p = 0.013$) in comparison to if tested in morning (Figure 7.14).

- There was a significant effect across within-subjects time point in Reasoning, $F(2,18) = 21.832$, $p < 0.017$. The subject's performance in reasoning was the best when tested in the morning in comparison to that of the performance in the afternoon ($p = 0.000$) and evening ($p = 0.004$). There was no difference in the student's performance on reasoning tested in the afternoon and evening (Figure 7.15).

Inference

On the basis of the sampled data in the illustration, it may be concluded that on the whole the academic performance of the students varies if tested at different time of the day. In Mathematics performance of the students was better when tested either in the afternoon or in the evening. The performance in English was the best if tested in the afternoon, whereas in Reasoning the performance of the students was best if tested in the morning instead of afternoon or in the evening.

EXERCISE

7.1.   Explain the situations where repeated measures MANOVA can be used.

7.2.   Discuss the assumptions of one-way repeated measures MANOVA. What course of action you will take if dependent variables are not related or when the multicollinearity exists.

7.3.   Consider an experiment where one-way repeated measures MANOVA is used to compare three different middle segment cars of Honda, Toyota, and Ford companies. Each of the six subjects in the study rate these cars on the issue of how much they are spacious. If space characteristics are assessed by means of three variables, leg space, roof top, and dickey, then show the layout design in the study.

7.4. Discuss the steps involved in solving a one-way repeated measure MANOVA design.

7.5. Identify a study in which one-way repeated measures MANOVA can be applied and write the hypotheses which you would like to test. How the issue of inflating the family-wise error rate is taken care of in this design?

ASSIGNMENT

7.1. A mobile company wanted to have a feedback as to how its three different models of mobiles are received by the customers. Eight randomly selected subjects participated in the study. Each subject was asked to rate all the three mobiles in a random order on its three characteristics; look, features, and security. The data so obtained are shown in Table 7.10. Apply one-way repeated measures MANOVA to investigate the following:
   a. Which mobile is liked by the subjects on the whole?
   b. Which is the most attractive characteristics to the subjects?
   c. Test your hypothesis by taking family-wise error rate (α) as 0.05.

7.2. A health scientist organized an experiment to study the changes in performance parameters on school children during lifestyle practices intervention. Each subject was tested for the concentration, memory, and alertness before starting the intervention and after two and four weeks of the intervention. The data so obtained are shown in Table 7.11. By assuming family-wise error rate (α) as 0.05, apply one-way repeated measures MANOVA to investigate the following:
   a. Which duration is better to improve performance of students on the whole?
   b. Which duration is better to improve each performance parameter?

Table 7.10 Ratings of the Subjects on Different Characteristics of Mobiles

Mobile A			Mobile B			Mobile C		
Look	Feature	Security	Look	Feature	Security	Look	Feature	Security
23	22	25	25	24	20	27	23	22
23	24	27	26	25	23	26	24	20
25	19	26	29	25	24	26	25	24
23	18	24	25	26	23	25	26	23
24	17	27	24	24	20	24	22	25
24	21	25	29	26	21	26	26	26
25	20	24	26	25	19	25	24	20
22	23	24	25	22	18	24	23	24
23	21	24	26	25	22	23	25	21
25	20	25	27	27	21	26	24	23

Table 7.11 Data on Performance Parameters of the Subjects at Different Duration

	Zero Day			Two Weeks			Four Weeks		
Concentration	Memory	Alertness		Concentration	Memory	Alertness		Concentration	Memory	Alertness
34	32	35		28	34	34		27	33	38
32	33	36		30	38	37		26	37	41
29	31	34		31	36	33		22	35	37
28	32	33		24	38	35		25	39	37
31	25	35		28	33	34		25	34	38
30	24	36		36	30	38		30	32	37
32	24	35		34	25	32		28	27	36
33	26	32		30	32	33		29	30	38
27	20	35		23	27	32		21	26	36
26	28	30		22	29	33		23	30	37

BIBLIOGRAPHY

Carey G. (1998). Multivariate analysis of variance (MANOVA): I. Theory. Retrieved May, 14, 2011.

Francis G. *Introduction to SPSS for Windows: v. 15.0 and 14.0 with Notes for Studentware.* 5th ed. Sydney: Pearson Education; 2007. Section 5.3.

Green SB, Salkind NJ, Akey TM. *Using SPSS for Windows and Macintosh: Analyzing and Understanding Data.* New Jersey: Prentice Hall; 2008.

Grimm LG, Yarnold PR, editors. *Reading and Understanding Multivariate Statistics.* Washington, D.C.: American Psychological Association; 1995.

Hair JF, Anderson RE, Tatham RL, Black WC. *Multivariate Data Analysis.* 5th ed. New York: Macmillan; 1998. (Chapter 6).

Hardle W, Simar L. *Applied Multivariate Statistical Analysis.* Vol. 289–303. Berlin, Heidelberg: Springer; 2007.

Huberty CJ, Olejnik S. *Applied MANOVA and Discriminant Analysis.* 2nd ed. Hoboken, New Jersey: John Wiley and Sons, Inc.; 2006.

Krzanowski WJ. *Principles of Multivariate Analysis: A User's Perspective.* UK: Oxford University Press; 2000.

Pallant J. *SPSS Survival Manual: A Step by Step Guide to Data Analysis Using SPSS for Windows (Versions 12–14).* 2005.

Stevens JP. *Applied Multivariate Statistics for the Social Sciences.* 4th ed. Mahwah, New Jersey: Lawrence Erlbaum Associates, Inc.; 2002.

Tabachnick BG, Fidell LS. *Using Multivariate Statistics.* 5th ed. Boston: Pearson International Edition; 2007. ISBN: 978-0-205-45938-4.

Tatsuoka MM. *Multivariate Analysis: Techniques for Educational and Psychological Research.* New York: John Wiley and Sons; 1971.

Timm NH. *"The General MANOVA Model (GMANOVA)" (Chapter 3.6.d)". Applied Multivariate Analysis. Springer Texts in Statistics.* New York: Springer-Verlag; 2002. ISBN: 0-387-95347-7.

8

MIXED DESIGN WITH TWO-WAY MANOVA

INTRODUCTION

In mixed design with two-way multivariate analysis of variance (MANOVA), effect of two factors (between-subjects and within-subjects) on a group of dependent variables is investigated simultaneously. This design is also known as split plot design. The mixed design with two-way MANOVA is particularly useful in a situation when group difference on a latent variable is required to be compared across different levels of the between-subjects as well as within-subjects factors. Variables like health, personality, aggression, quality of drinks are the examples of latent variable. These variables are known as latent variables because they are not directly measured. For instance, health can be assessed by the parameters like blood pressure, heart beat, and BMI; personality can be assessed by the constructs like openness, agreeableness, and conscientiousness; aggression can be measured by the dimensions like anger, hostility, and impulsivity; and quality of drinks may be assessed by the variables like sweetness, flavor, and hardness. The advantage of this design is that one can investigate the multivariate as well as univariate effects of within-subjects and between-subjects factors along with the interaction between them on a group of dependent variables. In mixed design with two-way MANOVA, the focus of an investigator is to see whether the multivariate effect across the interaction between within-subjects and between-subjects variables is significant or not. If the multivariate effect of interaction is significant, then comparing multivariate effect across the levels of the between-subjects variable and across the levels of the within-subjects variable

Repeated Measures Design for Empirical Researchers, First Edition. J. P. Verma.
© 2016 John Wiley & Sons, Inc. Published 2016 by John Wiley & Sons, Inc.

becomes meaningless. In that case univariate analysis is carried out for analyzing the interaction effect using mixed design with two-way ANOVA for each dependent variable separately. But if the multivariate interaction effect is not significant, the whole attention of the researcher is focused on investigating the multivariate main effects of between-subjects and within-subjects variables on the group of dependent variables.

Mixed design with two-way MANOVA can be planned for two types of research problems. Firstly, where levels of the within-subjects factor are different treatments and secondly, where its levels are different time durations. Consider a study in which the effect of hypertension and caffeine is investigated on the aggression. Here hypertension is between-subjects factor with two levels (hypertensive and non-hypertensive), whereas caffeine is a within-subjects factor with three different intensities (low, medium, and high) and aggression is a dependent variable, measured with three constructs like anger, hostility, and impulsivity. In this experiment levels of the within-subjects factor (caffeine) are different treatments. Here the main interest of the researcher is to compare the effects of different levels of caffeine on the aggression. Now look into another experiment where effects of gender and time on the fitness status are to be investigated during a 6-weeks exercise programme. Here gender is a between-subjects factor with two levels (male and female), time is a within-subjects variable with some levels say (initial, 4 weeks, 8 weeks, and 12 weeks), and fitness status is a dependent variable which may be measured through the three parameters (cardio, strength, and flexibility). The purpose of this experiment is to study the trend of improvement in fitness status of the subjects during the exercise programme. Another purpose might be to find the optimum duration for the improvement in fitness status during exercise intervention. In this chapter we shall discuss the analysis of the first type of design where the levels of the within-subjects factor are different treatments. This design shall be solved by using the SPSS.

WHAT HAPPENS IN MANOVA EXPERIMENT

One of the main advantages of the MANOVA analysis in mixed design is that it facilitates an investigator to analyze the multivariate as well as univariate effects. In other words, it helps a researcher to compare the means of identified groups of subjects on a combination of dependent variables as well as on each of the dependent variables. If multiple ANOVA experiments are planned instead of MANOVA, Type I error rate inflates. On the other hand, MANOVA experiment provides a procedure for controlling Type-I error because in such experiment univariate analysis is carried out only if the multivariate effect is significant. Another reason for MANOVA experiment to have more power is that it considers a set of different dependent variables as one single entity. This single entity is just like a super-variable or meta-variable. The next question is how to combine these dependent variables to obtain the meta-variable. MANOVA creates this meta-variable by using a linear combination of the dependent variables so as to maximize the group difference. The group difference on this meta-variable is tested for significance by using either of the two most popular multivariate tests Wilks' Lambda or Pillai's Trace. These tests are equivalent to F test in univariate ANOVA.

ASSUMPTIONS

In using mixed design with two-way MANOVA, certain assumptions need to be satisfied. MANOVA is very robust against the violation of normality assumption, but is very sensitive to the assumption of homogeneity of variance covariance matrices. The assumptions like data type, sample size, and independence of observations are the design issues. The SPSS software can be used to check most of these assumptions. Assumptions that are required for this design are as follows:

Multivariate Analysis

a. *Data type*. There should be two categorical independent variables, out of which one should be between-subjects and the other within-subjects. There should be two or more dependent variables measured on the metric scale. The dependent variables should be such that they can coherently be explained.

b. *Sample Size*. Number of observations must be at least higher than the number of dependent variables. It is recommended to have minimum sample size of 20 in this design.

c. *Independence of Observation*. The observations obtained on each subject must be independent to each other.

d. *Missing Data*. This design requires complete data for all subjects. In other words, no missing observation for any subject should be there in the experiment.

e. *Outliers*. No outlier should exist in each group of the independent variable for any of the dependent variables. This assumption can be verified by using the Box Plot generated in the SPSS.

f. *Linear Relationship*. All the dependent variables should be reasonably related to each other in each cell. These relationships should be linear. These assumptions can be tested by means of the bi-variate correlations and scatter plots generated by using SPSS.

g. *Normality*. Multivariate normality should exist. As an approximation it is good enough to test the normality of data in each cell. In other words, the data for each dependent variable in each level of the independent variable should be normal. Normality can be tested by using the Shapiro–Wilk test in SPSS. If normality assumption is violated, the procedure used for testing this design is robust enough to protect type I error but in that case power of the test is sacrificed.

h. *Multicollinearity*. There should be reasonable correlation among the dependent variables, but multicollinearity should not exist. If the correlation among dependent variables is 0.9 or more, multicollinearity exists and in that case, this assumption would be violated.

i. *Homogeneity of Variance Covariance Matrices*. The variance covariance matrices in all the levels of the between-subjects factor must be same. This assumption is tested by means of the Box's M test in SPSS. Since this assumption is very sensitive and affect the findings of the analysis too much, the significance of the Box's M test is tested at the level 0.001.

Univariate Analysis

j. *Sphericity*. There should be no sphericity in the data. If the sphericity assumption is violated, appropriate correction may be applied to compensate the type I error. This assumption can be tested by means of Mauchly's W test in SPSS. The sphericity assumption is tested only in a situation where the independent factor has more than two levels.

k. *Homogeneity of Variances*. Variance for the data obtained on each dependent variable must be same in all the levels of the between-subjects variable separately in each level of the within-subjects variable. This assumption is tested by the Levene's test.

LAYOUT DESIGN

Layout of the mixed design with two-way MANOVA depends upon whether the levels of the independent factor are different treatment conditions or time periods. Thus, the two different layout designs are as follows:

Case I: When the Levels of Within-Subjects Factor are Different Treatments

In this design layout, all the subjects in each level of the between-subjects factor are tested on multiple dependent variables under each treatment of within-subjects factor.

Reasonable time gap is given between any two treatments in order to done away the learning or fatigue effect while undergoing treatment. In this design counterbalancing is done to remove the order effect if any. Consider the above-discussed experiment of caffeine in which six hypertensive and six non-hypertensive subjects are tested for their aggression through three constructs like anger, hostility, and impulsivity after exposing them to three different levels (low, medium, and high) of caffeine consumption. One of the layouts in the mixed design with two-way MANOVA in this experiment can be shown by the Figure 8.1.

In this design all subjects in each level of between-subjects factor (hypertensive status) are tested on all the three dependent variables (anger, hostility, and impulsivity) in each level of the within-subjects factor (caffeine), but not in a particular sequence. All the six hypertensive subjects are divided into three groups randomly and then treatments are randomly allocated to these groups. In may be seen from the testing protocol in Figure 8.1 that in the first phase of testing the hypertensive subjects, H2 and H5, are tested on all the three dependent variables after exposing them to low caffeine, H3 and H6 are tested after exposing them to medium caffeine, and H1 and H4 are tested after having being exposed to high caffeine. This order has been randomized in the second and third phase of testing. Randomization of allocation of treatments to the subjects can be done differently also, the only consideration is that all the subjects in each category of the between-subjects variable should be tested in all the three treatment conditions. Similarly, randomization of treatments can be done for the subjects in the non-hypertensive group as shown in the Figure 8.1.

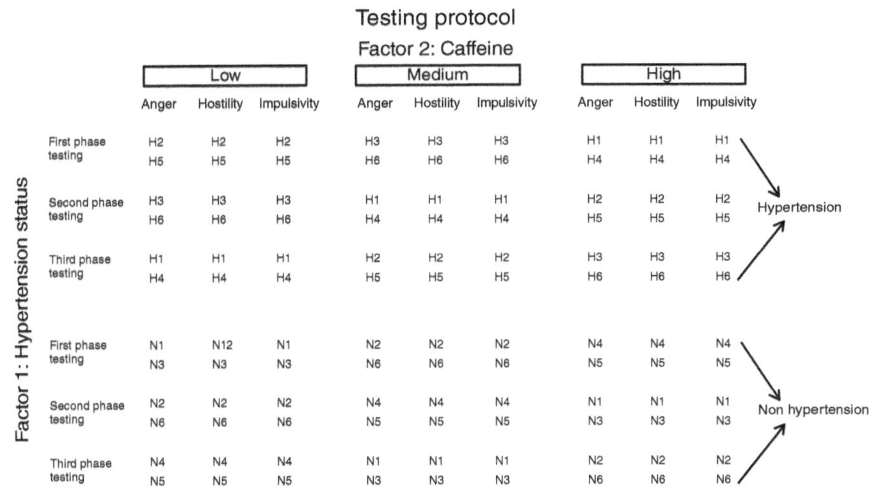

Testing protocol

Factor 2: Caffeine

		Low			Medium			High	
	Anger	Hostility	Impulsivity	Anger	Hostility	Impulsivity	Anger	Hostility	Impulsivity
First phase testing	H2 H5	H2 H5	H2 H5	H3 H6	H3 H6	H3 H6	H1 H4	H1 H4	H1 H4
Second phase testing	H3 H6	H3 H6	H3 H6	H1 H4	H1 H4	H1 H4	H2 H5	H2 H5	H2 H5
Third phase testing	H1 H4	H1 H4	H1 H4	H2 H5	H2 H5	H2 H5	H3 H6	H3 H6	H3 H6
First phase testing	N1 N3	N12 N3	N1 N3	N2 N6	N2 N6	N2 N6	N4 N5	N4 N5	N4 N5
Second phase testing	N2 N6	N2 N6	N2 N6	N4 N5	N4 N5	N4 N5	N1 N3	N1 N3	N1 N3
Third phase testing	N4 N5	N4 N5	N4 N5	N1 N3	N1 N3	N1 N3	N2 N6	N2 N6	N2 N6

Factor 1: Hypertension status

Hypertension

Non hypertension

Figure 8.1 Layout of the mixed design with two-way MANOVA where levels of the within-subject factor are different treatment conditions

Case II: When the Levels of the Within-Subjects Factor are Different Time Durations

The purpose of this design is to investigate the response pattern of the subjects on a group of dependent variables in different durations of the within-subjects factor (Time) during treatment. The randomization and counterbalancing are non-issues in this design. The researcher is mainly concerned in studying the trend of improvement in different durations and to compare such trends in different levels of the between-subjects factor. Consider the above-discussed example of studying the effect of exercise programme on fitness parameters. Let us assume that the six male and six female subjects are randomly selected for the study to see the impact of the exercise programme on their fitness status (cardio, strength, and flexibility) in different durations of the training (initial, 4 weeks, 8 weeks, and 12 weeks). In this design all the subjects in each category of the between-subjects (gender) factor are tested on all the dependent variables (cardio, strength, and flexibility) repeatedly at each time point (different levels of within-subjects factor). One of the layouts of this type of mixed design with two-way MANOVA can be shown by the Figure 8.2.

APPLICATION

There may be lot of application of the mixed design with two-way MANOVA in different disciplines. Some of the specific applications are discussed as follows:

1. A medical researcher may like to see the response of tuberculosis drug on the conditions of the male and female patients over the period of time during the treatment. In doing so a mixed design with two-way MANOVA experiment may

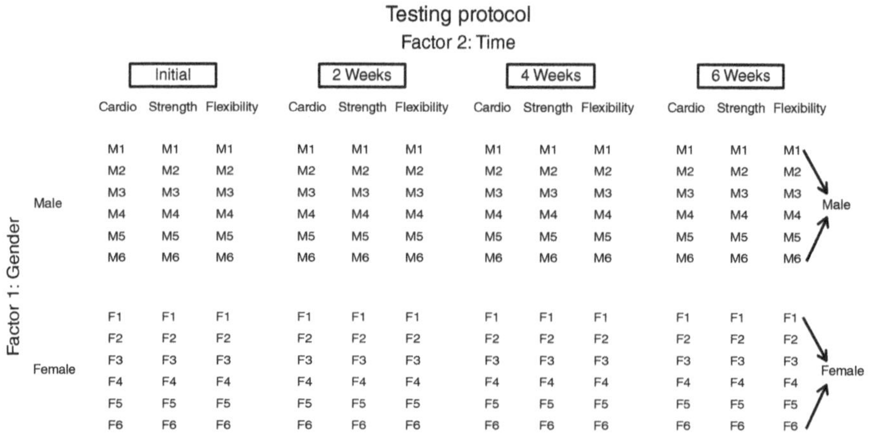

Testing protocol
Factor 2: Time

	Initial			2 Weeks			4 Weeks			6 Weeks		
	Cardio	Strength	Flexibility	Cardio	Strength	Flexibility	Cardio	Strength	Flexibility	Cardio	Strength	Flexibility
Male	M1	M1	M1	M1	M1	M1	M1	M1	M1	M1	M1	M1
	M2	M2	M2	M2	M2	M2	M2	M2	M2	M2	M2	M2
	M3	M3	M3	M3	M3	M3	M3	M3	M3	M3	M3	M3
	M4	M4	M4	M4	M4	M4	M4	M4	M4	M4	M4	M4
	M5	M5	M5	M5	M5	M5	M5	M5	M5	M5	M5	M5
	M6	M6	M6	M6	M6	M6	M6	M6	M6	M6	M6	M6
Female	F1	F1	F1	F1	F1	F1	F1	F1	F1	F1	F1	F1
	F2	F2	F2	F2	F2	F2	F2	F2	F2	F2	F2	F2
	F3	F3	F3	F3	F3	F3	F3	F3	F3	F3	F3	F3
	F4	F4	F4	F4	F4	F4	F4	F4	F4	F4	F4	F4
	F5	F5	F5	F5	F5	F5	F5	F5	F5	F5	F5	F5
	F6	F6	F6	F6	F6	F6	F6	F6	F6	F6	F6	F6

Factor 1: Gender (Male / Female)

Figure 8.2 Layout of the mixed design with two-way MANOVA where levels of the within-subjects factor are time durations

be planned. The patients in the male and female categories may be tested on the parameters like serum LDH, lungs power, and weight before starting the drug, after two, four, and six weeks of the drug administration. Here time would be a within-subjects variable, whereas gender would be a between-subjects variable.

2. A market researcher may wish to investigate the effect of gender and toothpaste brand on the buying behavior of customers on the basis of toothpaste features (therapeutic, taste, and fragrance). This analysis may reveal the difference in mean response on the features across different brands of toothpaste. Further, in-depth analysis may reveal as to which feature is more useful in buying a particular brand of toothpaste and the preference of gender for a particular feature of toothpaste based on a particular brand.

3. A nutritionist may wish to investigate the effect of gender and duration on the change in lifestyle indicators (fat%, cholesterol, and weight) in a six-week health awareness programme. This design facilitates a researcher to investigate performance trend of subjects during training programme. The subjects in male and female categories may be tested repeatedly on fat%, cholesterol, and weight before starting the programme and after two, four, and six weeks of the programme. Here the gender would be a between-subjects variable with two levels (male and female), whereas the time would be a within-subjects variable with four levels (initial, two weeks, four weeks, and six weeks).

STEPS IN SOLVING MIXED DESIGN WITH TWO-WAY MANOVA

In mixed MANOVA analysis is done in two phases. Firstly, the effect of independent variables is examined on the combined group of dependent variables, and secondly, the effects of independent variables are investigated in each of the dependent variable separately. While group differences (of dependent variables) in different categories of the independent variable are checked with multivariate techniques using MANOVA, the individual differences (each dependent variable) are assessed by using the uni-variate ANOVA. Following steps are used in solving this design.

1. Check the assumptions of data type, independence of measurements, outliers, linearity, normality, multicollinearity, equality of variance, and equality of variance–covariance matrices.
2. Describe layout of the design
3. Specify research questions to be investigated
4. Formulate multivariate and univariate hypotheses to be tested
5. Decide the family-wise error rate (α) for the design.
6. Use SPSS commands to generate the following outputs
   a. Levene's test for equality of variances
   b. Box's M test for equality of variance covariance matrices
   c. MANOVA table containing Wilks' Lambda and other multivariate test statistics for testing the significance of between-subjects effect, within-subjects effect, and interaction effect between the two independent factors, on the group of dependent variables.
   d. Mauchly's test of sphericity for within-subjects variance in each of the dependent variable (all levels of the between-subjects combined)
   e. ANOVA table for testing the significance of between-subjects variable on each of the dependent variable (all levels of within-subjects variables combined).
   f. Estimated marginal means for between-subjects main effect comparisons
   g. Repeated measures ANOVA table for testing the significance of within-subjects factor (all levels of between-subjects factor combined) and interaction effect on each of the dependent variable
   h. Estimated marginal means for within-subjects main effect comparisons.
   i. Marginal means plots
7. If interaction effect in the MANOVA table generated in the output 6(c) is not significant, report the effect of between-subjects and within-subjects factors in the table and conclude accordingly. In that situation perform repeated mixed rANOVA for each dependent variable separately to investigate main effects. But if the interaction is significant, find the simple effect of between-subjects and within-subjects factors for each dependent variable separately.
8. Simple effect of within-subjects factor is obtained by using the SPSS commands for the one-way repeated measures ANOVA after splitting the data file (developed in the MANOVA analysis). The following outputs are generated:
   a. Mauchly's Test of Sphericity
   b. *F*-tests for within-subjects effects in each category of the between-subjects factor
   c. Descriptive statistics for each dependent variable in each category of the between-subjects factor

   d. Pair-wise comparisons of marginal means in each category of the
      between-subjects factor
 9. Simple effect of between-subjects factor is obtained by using the SPSS com-
    mands for the independent measures one-way ANOVA. The following outputs
    are generated:
    a. Descriptive statistics
    b. Levene's test for equality of variances
    c. ANOVA table for testing the effect of between-subjects factor in each level
       of the within-subjects factor.
    d. Post hoc table for pair-wise comparison of means (if levels of the
       between-subjects are more than two).
10. Report the findings of MANOVA and univariate ANOVAs along with simple
    effects to address the research questions.

   The procedure used in solving this design shall be explained with the help of the
illustration discussed below.

ILLUSTRATION

A chocolate manufacturing company wishes to introduce three different types of
chocolate in the market. Before its launch, marketing research team wishes to know
as to how the people in the age category of 15 to 25 would react to these chocolates.
Further, they wish to investigate gender preference to different types of chocolates. A
study was planned in which 10 male and 10 female subjects were randomly selected.
They were given three types of chocolates (dark, milk, and white) to taste in a random
order after some interval of time. After they consumed the chocolate, they were asked
to give their response on the three characteristics of the chocolate (taste, crunchiness,
and flavor). The responses so obtained are shown in the Table 8.1. Let us see how
the two factors mixed MANOVA technique can be used to solve this between-within
design to address the issues of the research team.

Layout Design

In this illustration the effect of chocolate type and sex on the subject's response
towards chocolate characteristics shall be investigated. The response of all the male
and female subjects was taken after tasting all the three types of chocolates. Here
chocolate type is a within-subjects and sex is a between-subjects factors, respectively.
The design used in this illustration is a mixed design with two-way MANOVA, the
layout of which can be shown by the Figure 8.3.
   In this design response of all the subjects in male and female sections were
recorded on all the three dependent variables (taste, crunchiness, and flavor) for each
type of the chocolate, but not in a particular sequence. All ten male subjects were
divided into three groups randomly and then treatments were randomly allocated to
these groups. It can be seen from the testing protocol in Figure 8.3 that in the first

ILLUSTRATION 197

Table 8.1 Response on the Characteristics of Chocolate

	Subject	Dark Chocolate			Milk Chocolate			White Chocolate		
		Taste	Crunchiness	Flavor	Taste	Crunchiness	Flavor	Taste	Crunchiness	Flavor
Sex Male	1	5	4	5	7	6	6	5	5	6
	2	4	5	4	5	5	7	6	4	5
	3	6	5	6	7	6	7	5	5	4
	4	5	4	7	8	7	8	7	5	5
	5	4	5	6	6	8	7	5	6	6
	6	5	6	4	7	7	8	6	5	5
	7	4	5	6	7	6	8	6	6	5
	8	6	5	5	8	8	7	5	5	6
	9	7	5	6	7	7	8	5	4	5
	10	5	6	4	7	7	7	6	5	4
Female	1	7	6	7	4	5	6	7	5	5
	2	6	8	6	3	4	5	5	5	4
	3	8	7	6	3	3	5	8	4	5
	4	6	8	8	5	4	6	7	3	6
	5	5	9	6	4	4	5	5	5	6
	6	7	8	5	6	6	4	6	4	5
	7	7	9	8	6	6	5	6	3	6
	8	5	9	6	5	8	6	5	4	5
	9	6	7	5	3	6	4	7	4	5
	10	8	7	6	4	4	5	4	5	6

Testing protocol

Factor 2: Chocolate

		Dark			Milk			White			
		Taste	Crunchiness	Flavour	Taste	Crunchiness	Flavour	Taste	Crunchiness	Flavour	
First phase testing		M1 M4 M8	M1 M4 M8	M1 M4 M8	M3 M5 M9	M3 M5 M9	M3 M5 M9	M2 M6 M7 M10	M2 M6 M7 M10	M2 M6 M7 M10	
Second phase testing		M3 M5 M9	M3 M5 M9	M3 M5 M9	M2 M6 M7 M10	M2 M6 M7 M10	M2 M6 M7 M10	M1 M4 M8	M1 M4 M8	M1 M4 M8	Male
Third phase testing		M2 M6 M7 M10	M2 M6 M7 M10	M2 M6 M7 M10	M1 M4 M8	M1 M4 M8	M1 M4 M8	M3 M5 M9	M3 M5 M9	M3 M5 M9	
First phase testing		F2 F5 S9	F2 F5 F9	F2 F5 F9	F1 F3 F8 F10	F1 F3 F8 F10	F1 F3 F8 F10	F4 F6 F7	S4 F6 F7	F4 F6 F7	Female
Second phase testing		F1 F3 F8 F10	F1 F3 F8 F10	F1 F3 F8 F10	F4 F6 F7	F4 F6 F7	F4 F6 F7	F2 F5 F9	F2 F5 F9	F2 F5 F9	
Third phase testing		S2 S6 S8	S2 S6 S8	S2 S6 S8	F2 F5 F9	F2 F5 F9	F2 F5 F9	F1 F3 F8 F10	F1 F3 F8 F10	F1 F3 F8 F10	

Factor 1: Sex

Figure 8.3 Layout of the mixed design with two factors in the illustration

phase of testing three male subjects, M1, M4, and M8 were asked for their response on the dark chocolate, M3, M5, and M9 on the milk chocolate, and M2, M6, M7, and M10 on the white chocolate. This order was randomized in the second and third phase of testing. Sufficient time gap was provided between different phases of testing in order to done away the carryover effect. Random allocation of treatments to the subjects was done for counterbalancing. Randomization of allocation of treatment conditions to the subjects can be done differently also; the only consideration is that each of the ten male subjects should be tested in all the three treatment conditions. Similarly randomization of treatment was done for the female subjects as shown in Figure 8.3.

Research Questions

The following research questions need to be investigated in this illustration.

a. Whether the chocolate type affects the subject's response on the overall chocolate characteristics irrespective of the sex?
b. Whether sex affects the subject's response on the overall chocolate characteristics irrespective of the chocolate types?
c. Whether interaction of sex and chocolate type affects the subject's response on the overall chocolate characteristics?
d. Whether the chocolate type affects the subject's response on each of the chocolate characteristics in each sex?
e. Whether the male and female response differs on each of the chocolate characteristics in each type of chocolate.

Hypotheses Construction

a. The first research question shall be tested by the following null hypothesis:

   H_0 : There is no difference between group mean vectors of the subject's response in three types of chocolate irrespective of the sex.

   against

   H_1 : At least one group mean vector differs.   (8.1)

   Mathematically the null hypothesis can be written as follows:

   $$H_0 : \begin{bmatrix} \mu_{Taste} \\ \mu_{Crunchiness} \\ \mu_{Flavour} \end{bmatrix}_{Dark\_Chocolate} = \begin{bmatrix} \mu_{Tastes} \\ \mu_{Crunchiness} \\ \mu_{Flavour} \end{bmatrix}_{Milk\_Chocolate}$$

   $$= \begin{bmatrix} \mu_{Taste} \\ \mu_{Crunchiness} \\ \mu_{Flavour} \end{bmatrix}_{Whit\_Chocolate}$$

ILLUSTRATION   **199**

b. The second research question shall be tested by the following null hypothesis:

H_0 : There is no difference between group mean vectors of the subject's response in two different sexes irrespective of the chocolate.

against

H_1 : At least one group mean vector differs.   (8.2)

Mathematically the null hypothesis can be written as follows:

$$H_0 : \begin{bmatrix} \mu_{\text{Taste}} \\ \mu_{\text{Crunchiness}} \\ \mu_{\text{Flavour}} \end{bmatrix}_{\text{Male}} = \begin{bmatrix} \mu_{\text{Tastes}} \\ \mu_{\text{Crunchiness}} \\ \mu_{\text{Flavour}} \end{bmatrix}_{\text{Female}}$$

c. The third research question shall be tested by the following null hypothesis

H_0 : There is no interaction between sex and chocolate type on group mean vectors of the subject's response.

against

H_1 : The interaction between sex and chocolate type on group mean vectors of the subject's response is significant.   (8.3)

d. The fourth research question shall be investigated by testing the following null hypothesis for each chocolate characteristics in male and female group separately.

$$H_0 : \mu_{\text{Dark\_Chocolate}} = \mu_{\text{Milk\_Chocolate}} = \mu_{\text{White\_Chocolate}}$$

against

H_1 : At least any one group mean differs.   (8.4)

e. The fifth research question shall be examined by testing the following null hypothesis for each chocolate characteristics in each chocolate type separately.

$$H_0 : \mu_{\text{Male}} = \mu_{\text{Female}}$$

against

$$H_1 : \mu_{\text{Male}} \neq \mu_{\text{Female}}$$ (8.5)

Testing the first set of hypotheses shall answer the first research questions as to whether the subject's response on the combined chocolate characteristics differs in three different chocolate groups. This hypothesis shall be tested by using the Wilks' test in MANOVA table.

Wilks' test is generally used in MANOVA analysis, but if the sample size is small or the levels of factor are two then Pillai's test proves to be the better option.

Since Wilks' lambda indicates the proportion of variance in the combination of dependent variables (taste, crunchiness, and flavor) that is unaccounted for by the independent variable (chocolate type); lesser its value more is the variation in the group mean vectors. Wilks' lambda statistic follows approximately F distribution, hence F statistic is also computed in the MANOVA output. Thus, for the maximum group difference Wilks' lambda should be low and the corresponding F value should be significant.

The second and third hypotheses shall also be tested by using the Wilks' lambda along with its associated F value computed for the Sex and Interaction in the MANOVA table. If the interaction is significant, the fourth and fifth set of hypotheses shall be tested by means of univariate analysis for each of the dependent variable separately. If F value in any univariate ANOVA for between-subjects factor is significant, the post hoc test shall be applied for comparing the group means using Tukey HSD test. Similarly if the F value in any univariate ANOVA for within-subjects factor is significant, the group means will be compared by using the Bonferroni correction.

Level of Significance

The family-wise error rate (α) shall be taken as 0.05 in this illustration. If the Wilks' test for interaction is significant, two univariate repeated measures ANOVA for within-subjects and three univariate independent measures ANOVA for between-subjects shall be applied in order to address research questions. This will inflate the family-wise error rate (α). To compensate this, appropriate correction in the level of significance would be made while testing the significance of F value in testing the simple effect.

Solving Mixed Design with Two-Way MANOVA Using SPSS

Procedure in solving the mixed design with MANOVA shall be discussed by means of the data shown in the illustration using SPSS. Reader should note carefully the format of data file in this design. In **Variable View** define all the below mentioned nine variables as metric:

Taste_Dark
Crunch_Dark
Flavor_Dark
Taste_Milk
Crunch_Milk
Flavor_Milk
Taste_White
Crunch_White
Flavor_White

In **Data View** enter the data of the Table 8.1 column-wise as shown in the Figure 8.4. Once the data file is prepared, click on the following sequence of

ILLUSTRATION 201

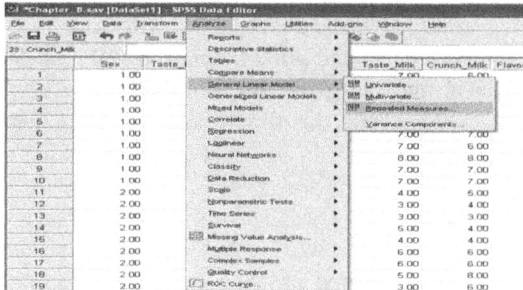

Figure 8.4 Data format in mixed design with two-way MANOVA

Figure 8.5 Screen for initiating commands for mixed design with MANOVA

commands as shown in Figure 8.5 to get the screen shown in Figure 8.6 for defining independent and dependent variables.

Analyze ⟶ General Linear Model ⟶ Repeated Measures

By default factor 1 is written in "Within-Subject Factor Name" section as shown in Figure 8.6a. Replace this by the independent variable Chocolate (within-subjects)

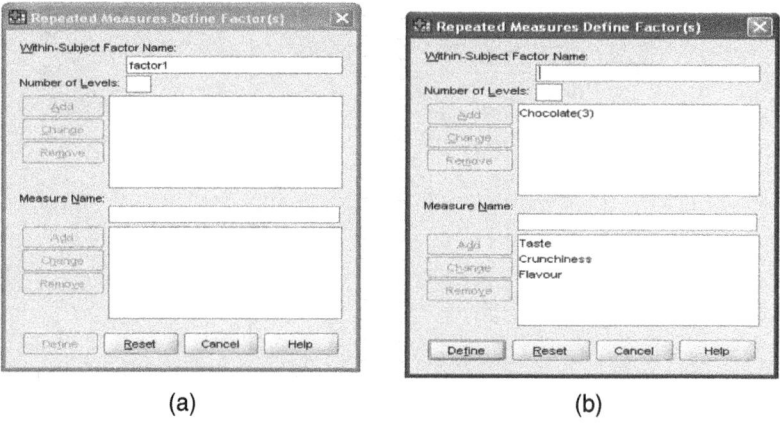

(a)   (b)

Figure 8.6 Screen showing options for defining independent and dependent variables

and write the number of levels as 3 as there are three levels (dark, milk, and white). Click on **Add** to enter this information into the box. Write the name of all the three dependent variables (Taste, Crunchiness, and Flavor) one by one in "Measure Name" section by clicking on **Add**. After defining these variables the screen shall look like as shown in Figure 8.6b. Be careful in defining the independent and dependent variables at this stage; the first letter should always be alphabet, and if the name consists of two words, it should be joined by using the underscore only. No other special character is permitted while defining the names of the variable.

After clicking on the command **Define** in the Figure 8.6b, the screen as shown in Figure 8.7a shall be obtained for selecting the variables defined in the data file. Selection of variable is very tricky at this stage.

Pickup the three variables *Taste_Dark, Taste_Milk*, and *Taste_White* from the list of variables in the left panel and bring them to the "Within-Subjects Variables" section of the screen in the right panel by using the arrow key. The screen shall look like Figure 8.7b. Now pick up the variables *Crunch_Dark, Crunch_Milk*, and *Crunch_White* and bring them in the right panel, thereafter shift the variables *Flavor_Dark, Flavor_Milk*, and *Flavor_White*. After shifting all the variables from the left panel to the right side the screen shall look like as shown in Figure 8.7c. Be careful about the order of the variables to be inserted in the section marked with "Within-Subjects Variables". Shifted variables should coincide with the sequence of the variables shown in the right panel in Figure 8.7a.

After selecting the variables, option needs to be defined for generating means plot in univariate repeated ANOVA for each dependent variable. Click on the **Plots** command to get the screen as shown in Figure 8.8. Do the following:

a. Transfer the variables *Sex* and *Chocolate* from the "Factors" section to the "Separate Lines" and "Horizontal Axis" areas, respectively, for generating means plots for each dependent variable for comparing the simple effects of between-subjects factor (Sex) in each level of the within-subjects factor (Chocolate). Click on **Add**.

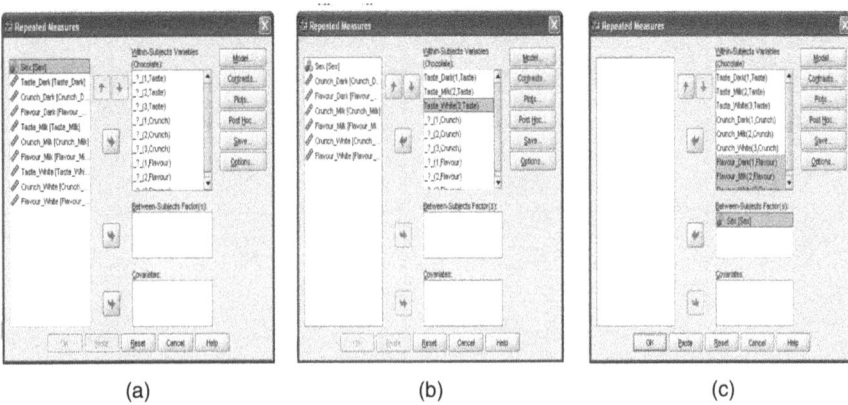

(a)   (b)   (c)

Figure 8.7   Screen showing option for selecting variables defining all treatment combinations

ILLUSTRATION   **203**

b. Transfer the variables Chocolate and Sex from the "Factors" section to the "Separate Lines" and "Horizontal Axis" areas, respectively, for generating means plots for each dependent variable for comparing the simple effect of within-subjects factor (Chocolate) in each level of the between-subjects factor (Sex). Click on **Add**.

After selecting the above-mentioned options for the various means plots click on **Continue** in the screen shown in Figure 8.8. This will take you back to the screen shown in Figure 8.7c.

Click on the option **Post Hoc** for testing the effect of between-subjects factor (Sex) in each level of the within-subjects factor (Chocolate) separately for each dependent variable. Transfer the variable *Sex* from the "Factor(s)" section to the "Post Hoc Tests for" section in screen as shown in Figure 8.9. Select 'Tukey' as an option for post hoc test. This is the most appropriate option for the post hoc test if the assumptions are satisfied. Click on **Continue** to get back the screen shown in Figure 8.7c for further options.

Click on **Options** command to open the Repeated Measures Options sub-dialogue box as shown in Figure 8.10. On this screen shift the factors *Sex, Chocolate,* and *Sex* × *Chocolate* from the "Factor(s) and Factor Interactions" section from the left panel to the "Display Means for" section in the right panel by using the arrow key. This will generate marginal means of Sex, Chocolate, and means for each cell.

Check the option 'Compare main effects' and select the option 'Bonferroni' in the dialogue box "Confidence Interval adjustment". This will generate the output for testing the effects of within-subjects factor (Sex) separately for each dependent variable. There is no post hoc test available for within-subjects factor; hence Bonferroni correction is required to compensate the inflated family-wise error rate (α) due to multiple comparisons. For detail discussion on this correction readers are advised to refer to the Chapter 3.

Check 'Descriptive statistics', 'Estimates of effect size', and 'Homogeneity tests' options in the "Display" area for generating various outputs related to descriptive

Figure 8.8 Screen showing option for various means plot for each dependent variable

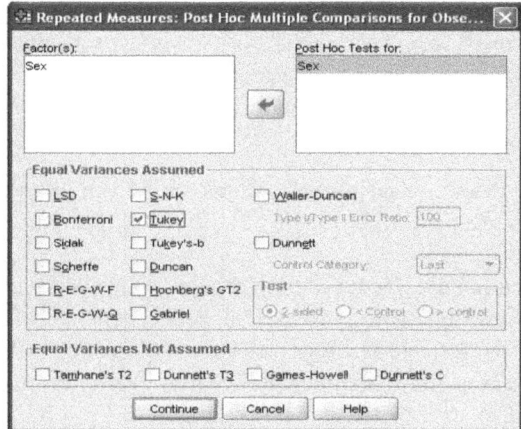

Figure 8.9 Screen showing option for post hoc test for the between-subjects factor (Sex) for each dependent variable

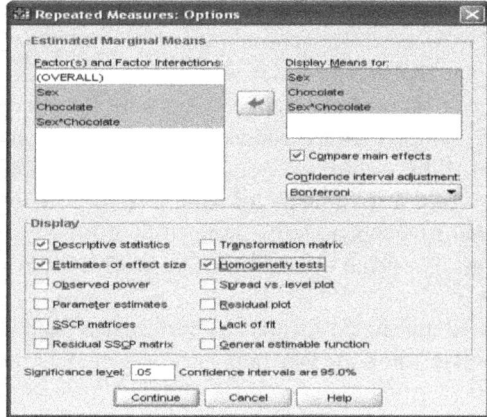

Figure 8.10 Screen showing options for comparing the effect of within-subjects factor (Chocolate) and other statistics

statistics, efficiency, and homogeneity conditions. Click on **Continue** to get back to the screen shown in Figure 8.7c. Click on **OK** for generating various outputs in this design.

SPSS Output and Interpretation

SPSS generates plenty of outputs in two-way mixed MANOVA, but only following relevant outputs need to be picked up for interpretation. Except marginal means plots, other outputs for the simple effect are not generated in this main analysis. These outputs are separately generated by using the procedure discussed later in this chapter.

ILLUSTRATION   **205**

Multivariate Outcome

1. Levene's test for equality of variances (Table 8.5)
2. Box's M test for equality of variance-covariance matrices (Table 8.6)
3. Multivariate tests of within-subjects effects (Table 8.8)

Main effect of Each Dependent Variable

4. Estimated marginal means for between-subjects main effect comparisons (Table 8.9)
5. Between group univariate ANOVA table for between-subjects main effect (Table 8.10)
6. Estimated marginal means for within-subjects main effect comparisons (Table 8.11)
7. Repeated measures ANOVA table for within-subjects main effect and interaction effect (Table 8.12)

Simple Effect of Each Dependent Variable

8. Mauchly's Test of Sphericity (Tables 8.13, 8.20, and 8.27)
9. *F*-tests for Chocolate (within-subjects) effects in each sex category (Tables 8.14, 8.21, and 8.28)
10. Descriptive statistics for each dependent variable in each sex (Tables 8.15, 8.22, and 8.29)
11. Pair-wise comparisons of marginal means in each sex category (Tables 8.16, 8.23, and 8.30)
12. Marginal means plot of each dependent variable (Sex × Chocolate) (Figures 8.17, 8.22, and 8.24)
13. Test of Homogeneity of Variances for each dependent variable (Tables 8.17, 8.24, and 8.31)
14. *F*-table for sex (between-subjects) effects in each chocolate category (Tables 8.18, 8.25, and 8.32)
15. Descriptive statistics for each dependent variable in each type of chocolate (Tables 8.19, 8.26, and 8.33)
16. Marginal means plot of each dependent variable (Chocolate × Sex) (Figures 8.21, 8.23, and 8.25)

Testing Assumptions

The mixed design with two-way MANOVA shall provide reliable findings only if the assumptions required for it hold true. Before discussing the findings of this analysis on the basis of the results generated in SPSS, let us first test these assumptions.

Data Type   In this illustration all the three dependent variables (taste, crunchiness, and flavor) are numeric and the independent variables (sex and chocolate) are categorical, hence the assumption of data type holds true.

Testing Correlations Utility of MANOVA design lies in the fact that the dependent variables are reasonably related with each other. Hence this assumption needs to be tested first. Let us examine the relationships among dependent variables in each category of the within-subjects factor by computing correlation matrix in male and female categories separately. This has been done by splitting the original data file (Figure 8.4) developed for the SPSS analysis.

Splitting Data File To split the data file into different sex category developed in SPSS, click on **Data** command in the header and then click on **Split File** in the submenu as shown in Figure 8.11. This will take you to the screen shown in Figure 8.12. Choose the radio button option 'Organize output by groups' and transfer the variable *Sex* from left panel into the area marked with "Grouped based on" in the right panel. Ensure that the radio button 'Sort the file by grouping variables' is selected. This option is selected by default. Click on **OK** to get the data file splitted as per the sex category.

After splitting the data file go back to the data file, click on the following sequence of commands and select all the variables except the sex in the window to generate the correlation matrix for male and female separately:

Analyze ⟶ Correlate ⟶ Bivariate

Tables 8.2 and 8.3 show the correlation matrix of all the variables in the male and female categories respectively.

In a situation where correlations among dependent variables are either very low or very high, the mixed design with two-way MANOVA is not appropriate. If the correlation is very low (<0.3), the univariate analysis should be done to investigate the effect of independent factors on each of the dependent variable separately instead of MANOVA. This yardstick is applicable only if the sample size in each cell is at least 20. However, if the correlation among dependent variables is very high (>0.9), it indicates multicollinearity and in that case one of the variables should be dropped because the variables having high correlation indicates the same thing.

Let us examine the relationship in this illustration. Table 8.2 shows that in the male category the correlation between Taste_Dark and Crunch_Dark is 0.000, and

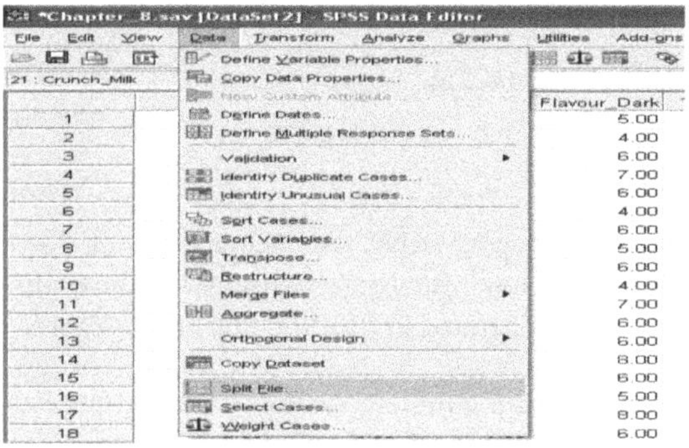

Figure 8.11 Screen for initiating commands for splitting data file

ILLUSTRATION   **207**

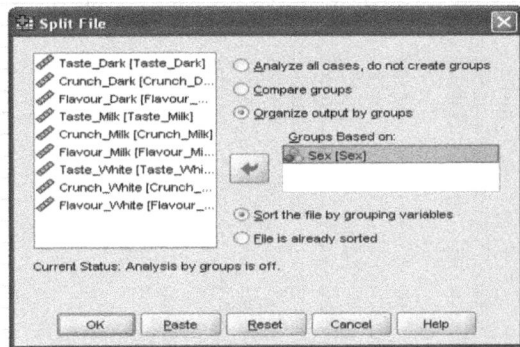

Figure 8.12   Screen showing option for splitting the data file in different sex category

that of between Taste_Dark and Flavor_Dark is 0.179. Similarly, the correlation between Taste_Milk and Flavor_Milk is 0.244, and that of between Crunch_Milk and Flavor_Milk is 0.156. In the white chocolate category, the correlation between Taste_White and Crunch_White is 0.000, and that of between Crunch_White and Flavor_White is 0.226.

On the other hand, Table 8.3 shows that in the female category the correlation between Taste_Dark and Flavor_Dark is 0.049, and that of between Crunch_Dark and Flavor_Dark is 0.162. In the milk chocolate category, the correlation between Taste_Milk and Flavor_Milk is 0.091, and that of between Crunch_Milk and Flavor_Milk is 0.000. On the other hand, in the white chocolate category the correlation between Taste_White and Flavor_White is −0.132.

All the correlations identified above indicate that the correlations among the dependent variables are quite low and do not make sense to use MANOVA analysis. But since these correlations have been calculated on the basis of only ten samples, it is not correct to draw any conclusion about the utility of the MANOVA analysis. If sample size would have been twenty or more in the illustration as recommended for this design, the correlation might have been significant provided theory supports clubbing these three variables together in assessing characteristics of chocolate.

Testing Normality   Before we proceed further we need to test the normality assumption. The SPSS provides the outputs for both the normality tests Kolmogorov–Smirnov as well as Shapiro–Wilk, but the Shapiro–Wilk test shall be retained here as the sample size is 10 only. By using the following sequence of commands and selecting the option 'Normality plots with test' after clicking on the **Options** command, the output shown in Table 8.4 has been obtained.

Analyze ⟶ Descriptive statistics ⟶ Explore

Readers are advised to refer to the Chapter 3 for its detail. Except Taste_Dark, Flavor_Dark, and Crunch_Milk, all other variables are not normally distributed in the male data. However, in the female data except Flavour_Milk, Crunch_White

Table 8.2 Correlations Matrix for Male Response

		Taste_Dark	Crunch_Dark	Flavor_Dark	Taste_Milk	Crunch_Milk	Flavor_Milk	Taste_White	Crunch_White	Flavor_White
Taste_Dark	Pearson correlation	1	0.000	0.179	0.523	0.271	0.116	-0.415	-0.503	-0.167
	Sig. (2-tailed)		1.000	0.620	0.121	0.449	0.750	0.232	0.139	0.646
	N	10	10	10	10	10	10	10	10	10
Crunch_Dark	Pearson correlation	0.000	1	-0.629	-0.190	0.176	0.247	0.000	0.000	-0.452
	Sig. (2-tailed)	1.000		0.051	0.598	0.627	0.492	1.000	1.000	0.190
	N	10	10	10	10	10	10	10	10	10
Flavor_Dark	Pearson correlation	0.179	-0.629	1	0.395	0.210	0.326	0.030	0.315	0.100
	Sig. (2-tailed)	0.620	0.051		0.258	0.560	0.357	0.934	0.376	0.784
	N	10	10	10	10	10	10	10	10	10
Taste_Milk	Pearson correlation	0.523	-0.190	0.395	1	0.495	0.244	0.109	0.190	0.017
	Sig. (2-tailed)	0.121	0.598	0.258		0.146	0.496	0.765	0.598	0.962
	N	10	10	10	10	10	10	10	10	10
Crunch_Milk	Pearson correlation	0.271	0.176	0.210	0.495	1	0.156	-0.201	0.351	0.365
	Sig. (2-tailed)	0.449	0.627	0.560	0.146		0.667	0.578	0.319	0.300
	N	10	10	10	10	10	10	10	10	10
Flavor_Milk	Pearson correlation	0.116	0.247	0.326	0.244	0.156	1	0.518	0.000	-0.290
	Sig. (2-tailed)	0.750	0.492	0.357	0.496	0.667		0.125	1.000	0.416
	N	10	10	10	10	10	10	10	10	10
Taste_White	Pearson correlation	-0.415	0.000	0.030	0.109	-0.201	0.518	1	0.000	-0.345
	Sig. (2-tailed)	0.232	1.000	0.934	0.765	0.578	0.125		1.000	0.330
	N	10	10	10	10	10	10	10	10	10
Crunch_White	Pearson correlation	-0.503	0.000	0.315	0.190	0.351	0.000	0.000	1	0.226
	Sig. (2-tailed)	0.139	1.000	0.376	0.598	0.319	1.000	1.000		0.530
	N	10	10	10	10	10	10	10	10	10
Flavor_White	Pearson correlation	-0.167	-0.452	0.100	0.017	0.365	-0.290	-0.345	0.226	1
	Sig. (2-tailed)	0.646	0.190	0.784	0.962	0.300	0.416	0.330	0.530	
	N	10	10	10	10	10	10	10	10	10

Bold face indicates that the effect is significant at 5% level.

Table 8.3 Correlations Matrix for Female Response

		Taste_Dark	Crunch_Dark	Flavor_Dark	Taste_Milk	Crunch_Milk	Flavor_Milk	Taste_White	Crunch_White
Taste_Dark	Pearson correlation	1	−0.598	0.049	−0.044	−0.414	−0.209	0.247	0.076
	Sig. (2-tailed)		0.068	0.894	0.903	0.234	0.562	0.491	0.834
	N	10	10	10	10	10	10	10	10
Crunch_Dark	Pearson correlation	−0.598	1	0.162	0.520	0.361	0.029	−0.431	0.255
	Sig. (2-tailed)	0.068		0.654	0.124	0.306	0.936	0.213	0.477
	N	10	10	10	10	10	10	10	10
Flavor_Dark	Pearson correlation	0.049	0.162	1	0.371	−0.141	0.668*	0.168	0.482
	Sig. (2-tailed)	0.894	0.654		0.291	0.698	0.035	0.642	0.159
	N	10	10	10	10	10	10	10	10
Taste_Milk	Pearson correlation	−0.044	0.520	0.371	1	0.514	0.091	−0.154	0.440
	Sig. (2-tailed)	0.903	0.124	0.291		0.128	0.803	0.672	0.203
	N	10	10	10	10	10	10	10	10
Crunch_Milk	Pearson correlation	−0.414	0.361	−0.141	0.514	1	0.000	−0.179	−0.110
	Sig. (2-tailed)	0.234	0.306	0.698	0.128		1.000	0.620	0.761
	N	10	10	10	10	10	10	10	10
Flavor_Milk	Pearson correlation	−0.209	0.029	0.668*	0.091	0.000	1	0.000	0.156
	Sig. (2-tailed)	0.562	0.936	0.035	0.803	1.000		1.000	0.667
	N	10	10	10	10	10	10	10	10
Taste_White	Pearson correlation	0.247	−0.431	0.168	−0.154	−0.179	0.000	1	−0.132
	Sig. (2-tailed)	0.491	0.213	0.642	0.672	0.620	1.000		0.716
	N	10	10	10	10	10	10	10	10
Crunch_White	Pearson correlation	0.000	−0.355	−0.479	−0.559	−0.283	−0.038	−0.452	−0.334
	Sig. (2-tailed)	1.000	0.315	0.162	0.093	0.427	0.917	0.190	0.346
	N	10	10	10	10	10	10	10	10
Flavor_White	Pearson correlation	0.076	0.255	0.482	0.440	−0.110	0.156	−0.132	−0.334
	Sig. (2-tailed)	0.834	0.477	0.159	0.203	0.761	0.667	0.716	0.346
	N	10	10	10	10	10	10	10	10

*Correlation is significant at the 0.05 level (2-tailed).

Table 8.4   Test of Normality

	Shapiro–Wilk (Male)			Shapiro–Wilk (Female)		
	Statistic	df	Sig.	Statistic	df	Sig.
Taste_Dark	0.886	10	0.152	0.907	10	0.258
Crunch_Dark	0.815	10	**0.022**	0.895	10	0.191
Flavor_Dark	0.874	10	0.111	0.846	10	0.051
Taste_Milk	0.82	10	**0.026**	0.878	10	0.124
Crunch_Milk	0.911	10	0.287	0.899	10	0.215
Flavor_Milk	0.802	10	**0.015**	0.833	10	**0.036**
Taste_White	0.781	10	**0.008**	0.94	10	0.55
Crunch_White	0.815	10	**0.022**	0.82	10	**0.025**
Flavor_White	0.833	10	**0.036**	0.802	10	**0.015**

Bold face indicates that the effect is significant at 5% level.

and Flavour_White variables all other variables are normally distributed. Since the MANOVA analysis is robust against the violation of normality, slight deviation from normality may not affect the findings severely. Only the variable Taste_White is severely non-normal, whereas all the identified variables are mildly non-normal. This can be compensated while testing the significance of the MANOVA statistic by decreasing the p value.

Testing Outliers   One of the main assumptions of this design is that there should be no outlier in the data of each cell. The Box plot shall be used to detect the outlier in the data. Since box plot for each variable in the male and female categories needs to be created, we shall workout with the same split file which was generated in computing correlation matrix. Click on the following commands in sequence:

Analyze ⟶ Descriptive statistics ⟶ Explore

Click on the command **Statistics** and select all the variables except Sex to create box plots for identifying outliers in male and female sections separately. Check the option 'Outliers' and click on **Continue** and **OK** to create various box plots.

Some of the box plots in the male category have been reported to have the outliers in this illustration. The box plot shown in Figures 8.13 reveals that only 1st and 6th data in the Crunch_Dark, 2nd, 4th, and 5th data in the Taste_Milk, and 2nd and 5th data in the Crunch_White indicated by the SPSS are the outliers. Since the data set is too small and the range of the response is also not large, it is up to the researcher to decide whether these data can be considered to be the outliers or not. In fact, SPSS only indicates that these data may be possible outliers; however, ultimate decision to recognize these data as outliers rests with the researcher. There was no outlier reported in the box plots generated for the data in the female category.

ILLUSTRATION   **211**

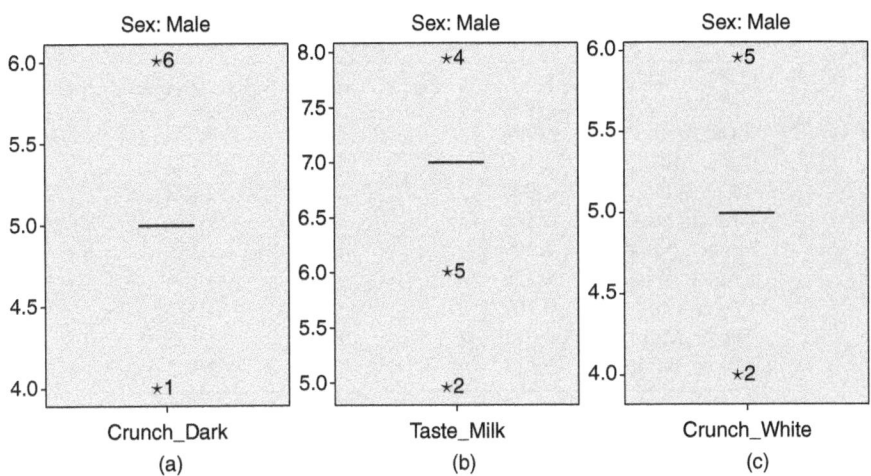

Figure 8.13   Box plots for the data in male category

Homogeneity of Variances   The outputs in Table 8.5 indicate that the assumption for between group homogeneity of variance across sex groups for all the three dependent variables (taste, crunchiness, and flavor), in each chocolate group, is satisfied as none of the F value is significant at 5% level ($p > 0.05$).

Homogeneity of Variance Covariance Matrices   The assumption of equality of variance covariance matrices is tested by means of the Box's M test. The output for this test is generated along with the output generated in this design in SPSS by selecting the appropriate option (Figure 8.10). The violation of this assumption severely affects the results in the MANOVA analysis, hence the significance of Box's M statistic is tested at the level 0.001. Since the F value associated with the Box's M test is not significant ($p > 0.001$), the assumption of homogeneity of variance covariance matrices is also met (Table 8.6).

Sphericity Assumption for Within-Subjects Conditions   The outcomes in the Table 8.7 are used for testing the sphericity assumption. Since the chi-square statistic associated with Mauchly's W statistic for each level of the within-subjects factor is not significant, we can state that the Sphericity is assumed. This will guide us to correct the line of univariate outcome in repeated measures ANOVA later.

Multivariate Testing

The first most important objective in using the mixed design with two-way MANOVA is to compare the group difference of the responses of the subjects on the combined characteristics of the chocolate in three different types of chocolate irrespective of the sex. The another important objective in this illustration is to know, whether response pattern of the male and female on the combined characteristics of the chocolate differs in three different types of chocolates. This would be tested by testing the multivariate

Table 8.5 Levene's Test of Equality of Error Variances[a]

	F	df1	df2	Sig. (p-value)
Taste_Dark	0.396	1	18	0.537
Taste_Milk	2.166	1	18	0.158
Taste_White	3.000	1	18	0.100
Crunch_Dark	3.524	1	18	0.077
Crunch_Milk	2.199	1	18	0.155
Crunch_White	1.328	1	18	0.264
Flavor_Dark	0.107	1	18	0.747
Flavor_Milk	0.012	1	18	0.913
Flavor_White	0.012	1	18	0.913

Tests the null hypothesis that the error variance of the dependent variable is equal across groups.
[a]Design: Intercept + Sex.
Within Subjects Design: Chocolate.

Table 8.6 Box's Test of Equality of Covariance Matrices[a]

Box's M	164.305
F	1.590
df1	45
df2	1.064E3
Sig. (p-value)	0.009

Tests the null hypothesis that the observed covariance matrices of the dependent variables are equal across groups.
[a]Design: Intercept + Sex.
Within Subjects Design: Chocolate.

Table 8.7 Mauchly's Test of Sphericity[b]

Within Subjects Effect	Measure	Mauchly's W	Approx. Chi-Square	df	Sig.	Epsilon[a]		
						Greenhouse–Geisser	Huynh–Feldt	Lower-bound
Chocolate	Taste	0.979	0.365	2	0.833	0.979	1.000	0.500
	Crunch	0.952	0.836	2	0.659	0.954	1.000	0.500
	Flavor	0.966	0.595	2	0.743	0.967	1.000	0.500

[a]May be used to adjust the degrees of freedom for the averaged tests of significance. Corrected tests are displayed in the Tests of Within-Subjects Effects table.
[b]Design: Intercept + Sex.
Tests the null hypothesis that the error covariance matrix of the orthonormalized transformed dependent variables is proportional to an identity matrix.
Within Subjects Design: Chocolate.

ILLUSTRATION 213

interaction effect. These two objectives shall be achieved by testing the hypotheses (8.1) and (8.3).

Mixed design with two-way MANOVA is in fact a factorial design and in any factorial design the primary focus of the researcher is to test the interaction effect, and if the interaction effect is significant, the simple effect is investigated. And in that case analyzing main effects of between-subjects and within-subjects factors becomes meaningless. However, if the interaction effect is not significant, only then it makes sense to test the main effects of the between-subjects and within-subjects factors.

Table 8.8 reveals MANOVA table for testing significance of the effect of Sex (between-subjects), Chocolate (within-subjects), and interaction between Sex and Chocolate on the group of dependent variables. Significance of the effect in MANOVA is usually tested by using the Wilks' Lambda or Pillai's Trace statistic. Both are powerful test but the researcher generally prefers Wilks' Lambda statistic for testing the effect. Other multivariate statistics that are used in testing the effect in MANOVA are Hotelling's Trace or Roy's Largest Root.

We shall use the Wilks' Lambda in this illustration to test significance of various effects. Since Wilks' lambda indicates the proportion of variance in the combined dependent variables that is unaccounted for by the independent variable, lesser its value more is the variation in the group mean vectors. Wilks' lambda statistic follows approximately F distribution; hence F statistic is also computed in the output of the MANOVA. Thus, for the maximum group difference Wilks' lambda should be low and the corresponding F value should be significant.

Table 8.8 shows that the multivariate effect of between-subjects factor (sex) on the combined characteristics of chocolate is not significant irrespective of the chocolate group: Wilks' $\lambda = 0.898$, $F(3,16) = 0.603$, $p = 0.623$; hence the null hypothesis (8.2) cannot be rejected. There is a significant multivariate effect of within-subjects factor (chocolate) on the combined characteristics of the chocolate irrespective of the sex: Wilks' $\lambda = 0.206$, $F(6,13) = 8.369$, $p = 0.001$. This indicates that the null hypothesis (8.1) is rejected and it may be concluded that there is a significant group difference in the subject's response on the combined dependent variables (taste, crunchiness, and flavor) among the three chocolate groups. There is also a significant multivariate effect across the interaction between the sex group and chocolate group: Wilks' $\lambda = 0.060$, $F(6,13) = 34.203$, $p = 0.000$. Thus, the null hypothesis (8.3) is rejected.

Since the multivariate effect across the interaction between the sex and chocolate groups is significant and also partial Eta Square is 0.940 which is considered to be very high, the univariate analysis of variance (ANOVAs) for each dependent variable shall be meaningful and would be of practical value. This analysis shall be applied as follow-up test of the MANOVA.

Univariate Testing

This mixed design with two-way ANOVA shall be solved for each dependent variable separately. The between-subjects main effect (sex) and within-subjects main effect (chocolate) shall be tested along with the interaction effect (sex × chocolate) for

Table 8.8 Multivariate Tests[b]

Effect			Value	F	Hypoth esis df	Error df	Sig.	Partial Eta Squared
Between Subjects	Intercept	Pillai's trace	0.997	1.656$E3$[a]	3.000	16.000	0.000	0.997
		Wilks' lambda	0.003	1.656$E3$[a]	3.000	16.000	0.000	0.997
		Hotelling's trace	310.456	1.656$E3$[a]	3.000	16.000	0.000	0.997
		Roy's largest root	310.456	1.656$E3$[a]	3.000	16.000	0.000	0.997
	Sex	Pillai's trace	0.102	0.603[a]	3.000	16.000	0.623	0.102
		Wilks' lambda	0.898	0.603[a]	3.000	16.000	0.623	0.102
		Hotelling's trace	0.113	0.603[a]	3.000	16.000	0.623	0.102
		Roy's largest root	0.113	0.603[a]	3.000	16.000	0.623	0.102
Within Subjects	Chocolate	Pillai's trace	0.794	8.369[a]	6.000	13.000	0.001	0.794
		Wilks' lambda	0.206	8.369[a]	6.000	13.000	0.001	0.794
		Hotelling's trace	3.863	8.369[a]	6.000	13.000	0.001	0.794
		Roy's largest root	3.863	8.369[a]	6.000	13.000	0.001	0.794
	Chocolate * Sex	Pillai's trace	0.940	34.203[a]	6.000	13.000	0.000	0.940
		Wilks' lambda	0.060	34.203[a]	6.000	13.000	0.000	0.940
		Hotelling's trace	15.786	34.203[a]	6.000	13.000	0.000	0.940
		Roy's largest root	15.786	34.203[a]	6.000	13.000	0.000	0.940

[a]Exact statistic.
[b]Design: Intercept + Sex.
Within Subjects Design: Chocolate.

ILLUSTRATION   **215**

each dependent variable (taste, crunchiness, and flavor) seperately. The family-wise error rate for the MANOVA analysis was fixed at 0.05. Now that the three mixed repeated measures ANOVA for each dependent variable shall be done for the same set of data as a follow-up for the MANOVA analysis, the family-wise error rate shall inflate, and to compensate this error, we shall test the univariate ANOVA by taking α as 0.017(0.05/3). Thus, the between-subjects (Sex) and within-subjects (Chocolate) main effects and interaction effect between them (Sex $\times$ Chocolate) shall be tested at the significance level (α) 0.017.

Main Effect of Between-Subjects Factor (Sex) Table 8.9 shows the estimated marginal means of each sex group in each of the three dependent variables. This information is needed to get the actual picture of the difference in the response of the subjects in both the sex groups (all chocolate groups combined) for each of the three dependent variable only if their associated F are significant.

Table 8.10 shows the ANOVA table for the Sex (between-subjects) in each of the three dependent variable. The result shows that the effect of Sex is not significant in any of the chocolate characteristics irrespective of the chocolate types (dark, milk, and white) because the p value associated with all the three F is more than 0.017.

Main Effect of Within-Subjects Factor (Chocolate) Table 8.11 indicates the estimated marginal means of each chocolate group in each of the three dependent variables. This information may be used to know the actual difference in the response of the subjects in the three chocolate groups (both sexes combined) for each of the dependent variable if their associated F is significant.

Table 8.12 shows the outputs in repeated measures ANOVA for testing the significance of the Chocolate (within-subjects) and the Interaction (Chocolate $\times$ Sex) effects on each of the three dependent variables (taste, crunchiness, and flavor) seperately. Since both these effects are repeated measures, checking sphericity assumption is important before drawing any conclusion. We have seen in Table 8.7 that the sphericity assumption has not been violated for any of the dependent variable; hence we can state that sphericity is assumed. This will guide us to the correct line of outcome in the ANOVA Table 8.12. This table shows that the effects of Chocolate on Crunchiness

Table 8.9 Estimates of Marginal Means (All Chocolate Groups Combined)

Measure	Sex	Mean	Std. Error	95% Confidence Interval	
				Lower Bound	Upper Bound
Taste	Male	5.867	0.196	5.456	6.278
	Female	5.600	0.196	5.189	6.011
Crunch	Male	5.567	0.186	5.176	5.958
	Female	5.667	0.186	5.276	6.058
Flavor	Male	5.900	0.190	5.502	6.298
	Female	5.567	0.190	5.168	5.965

Table 8.10   ANOVA Table for Testing Between-Subjects Effects in Each Dependent Variable

		Transformed Variable: Average					
Source	Measure	Type III Sum of Squares	df	Mean Square	F	Sig.	Partial Eta Squared
Intercept	Taste	1972.267	1	1972.267	1.718E3	0.000	0.990
	Crunch	1892.817	1	1892.817	1.822E3	0.000	0.990
	Flavor	1972.267	1	1972.267	1.830E3	0.000	0.990
Sex	Taste	1.067	1	1.067	0.929	0.348	0.049
	Crunch	0.150	1	0.150	0.144	0.708	0.008
	Flavor	1.667	1	1.667	1.546	0.230	0.079
Error	Taste	20.667	18	1.148			
	Crunch	18.700	18	1.039			
	Flavor	19.400	18	1.078			

Table 8.11   Estimates of Marginal Means (Both Sexes Combined)

Measure	Chocolate	Mean	Std. Error	95% Confidence Interval	
				Lower Bound	Upper Bound
Taste	Dark	5.800	0.232	5.312	6.288
	Milk	5.600	0.230	5.117	6.083
	White	5.800	0.226	5.325	6.275
Crunch	Dark	6.400	0.194	5.992	6.808
	Milk	5.850	0.279	5.263	6.437
	White	4.600	0.163	4.257	4.943
Flavor	Dark	5.800	0.237	5.302	6.298
	Milk	6.200	0.158	5.868	6.532
	White	5.200	0.158	4.868	5.532

($p = 0.000$) and Flavor ($p = 0.000$) are significant irrespective of the sexes because p values associated with their respective F are less than 0.017. However, no significant effect of Chocolate on taste has been observed.

Table 8.12 indicates that the interaction effect (chocolate $\times$ sex) is significant in each of the three dependent variables; Taste ($p = 0.000$), Crunchiness ($p = 0.000$), and Flavor ($p = 0.000$), as p values associated with their respective F are less than 0.017. Now that the interaction effect is significant in all the three dependent variables, the simple effects will be tested in each of these variables separately.

Remark: In this design we are mainly interested to know whether the interaction effect is significant. If it is so, a detail analysis is done to investigate the simple effect for each of the dependent variable separately. However, if the interaction effect is not significant, only the main effects of between-subjects and within-subjects factors are analyzed.

Table 8.12 Univariate Tests: Repeated Measures ANOVA Table for Each Dependent Variable

Source	Measure		Type III Sum of Squares	df	Mean Square	F	Sig. (p-value)	Partial Eta Squared
Chocolate	Taste	Sphericity assumed	0.533	2	0.267	0.266	0.768	0.015
		Greenhouse–Geisser	0.533	1.958	0.272	0.266	0.764	0.015
		Huynh–Feldt	0.533	2.000	0.267	0.266	0.768	0.015
		Lower-bound	0.533	1.000	0.533	0.266	0.613	0.015
	Crunchiness	Sphericity assumed	34.033	2	17.017	18.791	0.000	0.511
		Greenhouse–Geisser	34.033	1.908	17.833	18.791	0.000	0.511
		Huynh–Feldt	34.033	2.000	17.017	18.791	0.000	0.511
		Lower-bound	34.033	1.000	34.033	18.791	0.000	0.511
	Flavor	Sphericity assumed	10.133	2	5.067	9.702	0.000	0.350
		Greenhouse–Geisser	10.133	1.933	5.241	9.702	0.001	0.350
		Huynh–Feldt	10.133	2.000	5.067	9.702	0.000	0.350
		Lower-bound	10.133	1.000	10.133	9.702	0.006	0.350
Chocolate * Sex	Taste	Sphericity assumed	43.333	2	21.667	21.587	0.000	0.545
		Greenhouse–Geisser	43.333	1.958	22.127	21.587	0.000	0.545
		Huynh–Feldt	43.333	2.000	21.667	21.587	0.000	0.545
		Lower-bound	43.333	1.000	43.333	21.587	0.000	0.545
	Crunchiness	Sphericity assumed	56.700	2	28.350	31.307	0.000	0.635
		Greenhouse–Geisser	56.700	1.908	29.710	31.307	0.000	0.635
		Huynh–Feldt	56.700	2.000	28.350	31.307	0.000	0.635
		Lower-bound	56.700	1.000	56.700	31.307	0.000	0.635

(*continued*)

Table 8.12 *(Continued)*

Source	Measure		Type III Sum of Squares	df	Mean Square	F	Sig. (*p*-value)	Partial Eta Squared
	Flavor	Sphericity assumed	27.733	2	13.867	26.553	0.000	0.596
		Greenhouse–Geisser	27.733	1.933	14.344	26.553	0.000	0.596
		Huynh–Feldt	27.733	2.000	13.867	26.553	0.000	0.596
		Lower-bound	27.733	1.000	27.733	26.553	0.000	0.596
Error (Chocolate)	Taste	Sphericity assumed	36.133	36	1.004			
		Greenhouse–Geisser	36.133	35.251	1.025			
		Huynh–Feldt	36.133	36.000	1.004			
		Lower-bound	36.133	18.000	2.007			
	Crunchiness	Sphericity assumed	32.600	36	0.906			
		Greenhouse–Geisser	32.600	34.352	0.949			
		Huynh–Feldt	32.600	36.000	0.906			
		Lower-bound	32.600	18.000	1.811			
	Flavor	Sphericity assumed	18.800	36	0.522			
		Greenhouse–Geisser	18.800	34.803	0.540			
		Huynh–Feldt	18.800	36.000	0.522			
		Lower-bound	18.800	18.000	1.044			

ILLUSTRATION   **219**

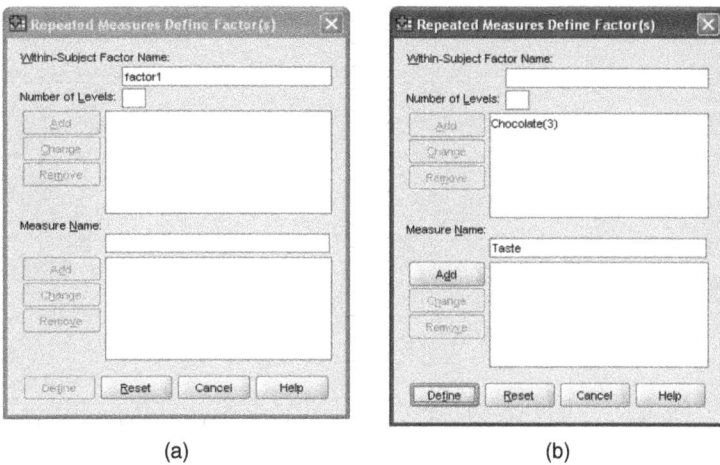

(a)   (b)

Figure 8.14 Screen showing options for defining variables

Level of Significance for Simple Effect In finding the simple effect of between-subjects variable (sex), three ANOVA tests shall be applied (one for each type of chocolate); hence the p value for testing the significance of F for sex in each type of chocolate shall be $0.0057 (= 0.017/3)$. Similarly in testing the significance of Chocolate two repeated measures ANOVA (one for each sex) shall be applied; hence for testing the F value for chocolate in each sex, the p value would be $0.0085 (= 0.017/2)$.

Simple Effect on Taste Simple effect on taste will give the correct picture about the response of the subjects in the three different types of chocolates on its taste in different sex groups. We need to investigate the simple effect of chocolate as well as simple effect of sex. Simple effect of chocolate will facilitate us to compare the response of the subjects on taste among three different types of chocolates in each sex group. Whereas, the simple effect of sex will provide us the comparison of the male and female response on the taste in each type of the chocolate separately.

Effect of Chocolate (Within-Subjects) The results of simple effect are not provided in SPSS outputs while analyzing the mixed design with two-way MANOVA. Some more work is required to be done to generate these outputs. To find the simple effect of Chocolate (within-subjects) for the data on taste, one-way repeated measures ANOVA needs to be performed in SPSS. Open the same data file as shown in Figure 8.4, split it as per the sex category the way we did in generating the correlation matrix. Click on the following commands in sequence to get the screen as shown in Figure 8.14a for defining independent and dependent variables.

Analyze $\longrightarrow$ General minear model $\longrightarrow$ Repeated measures

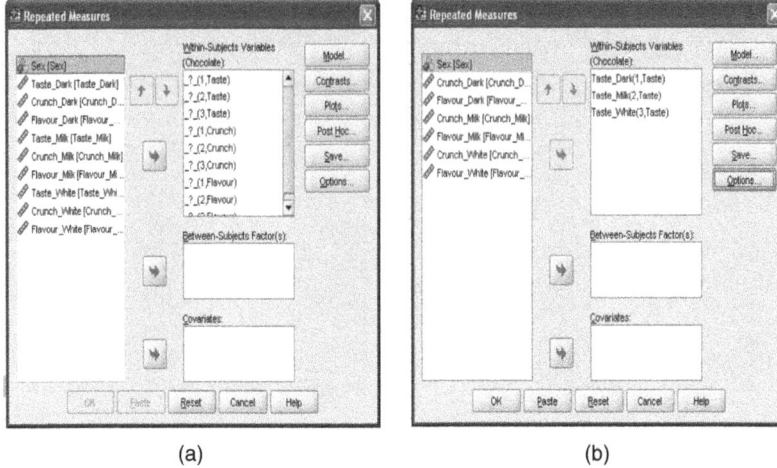

(a)   (b)

Figure 8.15   Screen showing option for selecting within-subjects variables

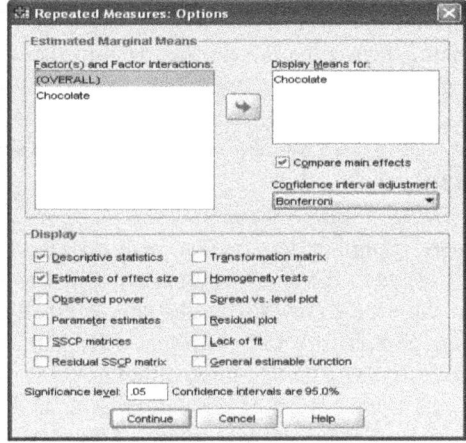

Figure 8.16   Screen showing options for generating the output for simple effect of within-subjects factor (Chocolate) and other statistics

Replace factor 1 by the independent variable Chocolate (within-subjects) and write the number of levels as 3. Click on **Add** to enter this information into the box. Write the name of the dependent variable for which the simple effect is required to be computed in "Measure Name" section. Here the variable shall be Taste. Click on **Add**.

After clicking on the command **Define** in the Figure 8.14b, the screen as shown in Figure 8.15a shall be obtained for selecting variables.

Pickup the variables *Taste_Dark*, *Taste_Milk*, and *Taste_White* from the list of variables in the left panel and bring them in the "Within-Subjects Variables" section of the screen in the right panel by using the arrow key. Click on **Options** command to open the Repeated Measures Options sub-dialogue box as shown in Figure 8.16.

ILLUSTRATION 221

Table 8.13 Mauchly's Test of Sphericity in Different Sex

						Epsilon		
Sex	Within Subjects Effect	Mauchly's W	Approx. Chi-Square	df	Sig. (*p*-value)	Greenhouse– Geisser	Huynh– Feldt	Lower- bound
Male	Chocolate	0.701	2.837	2	0.242	0.770	0.898	0.500
Female	Chocolate	0.922	0.648	2	0.723	0.928	1.000	0.500

Measure: **Taste**

Bring the factor Chocolate from the "Factor(s) and Factor Interactions" section in the left panel to the "Display Means for" section in the right panel by using arrow button. This will generate marginal means of Chocolate.

Check the option 'Compare main effects' and select the option 'Bonferroni' in the dialogue box "Confidence Interval adjustment". This will generate the output for testing the effects of within-subjects factor (Chocolate) separately for male and female. There is no post hoc test available for within-subjects factor; hence Bonferroni correction shall be used to compensate the inflated family-wise error rate (α) due to multiple comparisons. Click on **Continue** to get back to the screen shown in Figure 8.15b. Click on **OK** for generating various outputs for the simple effect. These commands shall generate the outputs as shown in the Tables 8.13 to 8.16 and Figure 8.17.

Table 8.13 shows that the chi-square statistic associated with Mauchly's W statistic for both the levels of the between-subjects factor, Sex, is not significant because *p* value associated with the respective chi-square is >0.05. Hence, we can state that the sphericity is assumed. This will guide us to correct the outcome in the repeated measures ANOVA in Table 8.14.

Table 8.14 indicates that the *F* values associated with the Chocolate (Sphericity assumed) in male ($p = 0.000$) and female ($p = 0.001$) sections are significant because *p* values associated with these *F* are less than 0.0085. This encourages us to make the pair-wise comparisons of marginal means in each sex category. Table 8.15 shows mean scores of taste for each chocolate category in male and female groups, whereas Table 8.16 shows the pair-wise comparison of these marginal means. By combining the contents of these two tables, marginal means plot of Sex × Chocolate for the data on Taste as shown in the Figure 8.17 can be obtained. In fact this means plot is generated in the outputs of the mixed design with two-way MANOVA in SPSS.

Means Plots (Sex × Chocolate) The means plot shown in Figure 8.17 can be used to compare the mean scores of taste in each type of chocolates in male and female groups separately. This means plot provides clear picture of the analysis. It indicates that the male subjects find milk chocolate to be significantly more tasty in comparison to that of white as well as dark chocolate. However, there is no difference in the taste of white and dark chocolate. On the other hand, female subjects find dark chocolate to be tastier than the milk chocolate.

Effect of Sex (Between-Subjects) To find the simple effect of Sex, it is required to compare the Taste scores of male and female in each category of chocolate

Table 8.14 *F*-Table for Testing Significance of Chocolate (Within-Subjects) Effects in Each Sex Category

	Measure: Taste						
Source		Type III SS	df	Mean Square	F	Sig. (*p*-value)	Partial Eta Squared
Sex: Male							
Chocolate	Sphericity assumed	17.267	2	8.633	12.878	0.000	0.589
	Greenhouse–Geisser	17.267	1.540	11.211	12.878	0.001	0.589
	Huynh–Feldt	17.267	1.797	9.611	12.878	0.001	0.589
	Lower-bound	17.267	1.000	17.267	12.878	0.006	0.589
Error (Chocolate)	Sphericity assumed	12.067	18	0.670			
	Greenhouse–Geisser	12.067	13.862	0.871			
	Huynh–Feldt	12.067	16.169	0.746			
	Lower-bound	12.067	9.000	1.341			
Sex: Female							
Chocolate	Sphericity assumed	26.600	2	13.300	9.947	0.001	0.525
	Greenhouse–Geisser	26.600	1.856	14.335	9.947	0.002	0.525
	Huynh–Feldt	26.600	2.000	13.300	9.947	0.001	0.525
	Lower-bound	26.600	1.000	26.600	9.947	0.012	0.525
Error (Chocolate)	Sphericity assumed	24.067	18	1.337			
	Greenhouse–Geisser	24.067	16.700	1.441			
	Huynh–Feldt	24.067	18.000	1.337			
	Lower-bound	24.067	9.000	2.674			

Table 8.15 Descriptive Statistics

	Mean	Std. Deviation	N
Sex: Male			
Taste_Dark	5.1000	0.99443	10
Taste_Milk	6.9000	0.87560	10
Taste_White	5.6000	0.69921	10
Sex: Female			
Taste_Dark	6.5000	1.08012	10
Taste_Milk	4.3000	1.15950	10
Taste_White	6.0000	1.24722	10

separately. In other words, three separate one-way independent measures ANOVA needs to be applied. This can be done by applying the SPSS commands for one-way ANOVA using the same data file shown in Figure 8.4, which we developed in the main MANOVA analysis. Click on the following sequence of commands in SPSS to get the screen as shown in Figure 8.18:

Analyze ⟶ Compare means ⟶ One-Way ANOVA

ILLUSTRATION

223

Table 8.16   Pair-wise Comparisons of Marginal Means in Each Sex Category

		Measure: Taste				
		Mean Diff			95% CI for Difference[a]	
Chocolate (I)	Chocolate (J)	(I − J)	Std. Error	Sig[a]	Lower Bound	Upper Bound
Sex: Male						
Dark	Milk	−1.800*	0.291	0.000	−2.652	−0.948
	White	−0.500	0.453	0.896	−1.830	0.830
Milk	Dark	1.800*	0.291	0.000	0.948	2.652
	White	1.300*	0.335	0.011	0.317	2.283
White	Dark	0.500	0.453	0.896	−0.830	1.830
	Milk	−1.300*	0.335	0.011	−2.283	−0.317
Sex: Female						
Dark	Milk	2.200*	0.512	0.006	0.698	3.702
	White	0.500	0.453	0.896	−0.830	1.830
Milk	Dark	−2.200*	0.512	0.006	−3.702	−0.698
	White	−1.700*	0.578	0.049	−3.396	−0.004
White	Dark	−0.500	0.453	0.896	−1.830	0.830
	Milk	1.700*	0.578	0.049	0.004	3.396

Based on estimated marginal means.
*The mean difference is significant at the 0.0085 level.
[a]Adjustment for multiple comparisons: Bonferroni.

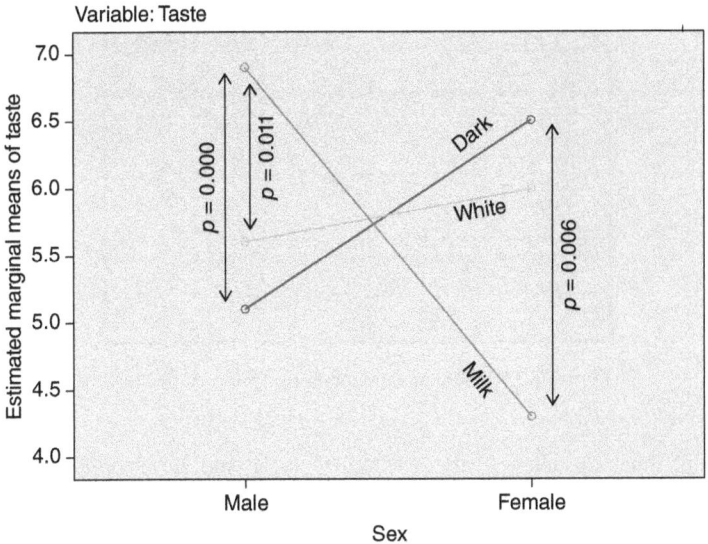

Figure 8.17   Marginal means plot of Sex × Chocolate for the data on Taste

Table 8.17   Test of Homogeneity of Variances

	Levene Statistic	df1	df2	Sig. (p-value)
Taste_Dark	0.396	1	18	0.537
Taste_Milk	2.166	1	18	0.158
Taste_White	3.000	1	18	0.100

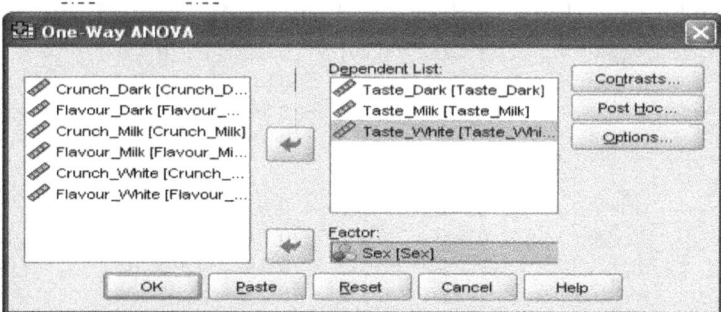

Figure 8.18   Screen showing option for selecting the variables for analyzing simple effect of Sex

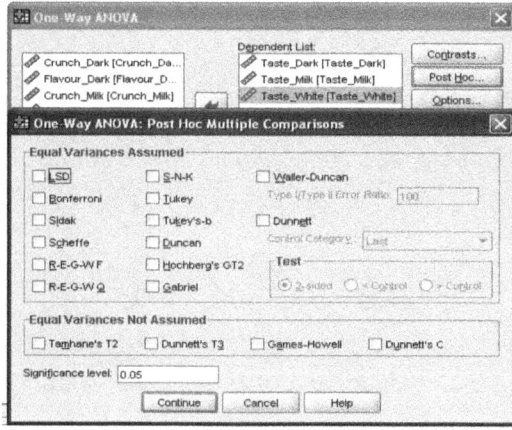

Figure 8.19   Screen showing option for post hoc test

Bring all three variables *Taste_Dark*, *Taste_Milk*, and *Taste_White* from the left panel into the "Dependent List" section of the right panel. Shift *Sex* variable from the left panel into the "Factor" area on the right panel of the window. Click on the **Post Hoc** and select the option 'Tukey' as shown in Figure 8.19. Click on **Continue** to go back to the screen as shown in Figure 8.18. Now click on the **Option** and check the option for 'Descriptive', 'Homogeneity of variance test', and 'Means plot' as

ILLUSTRATION   225

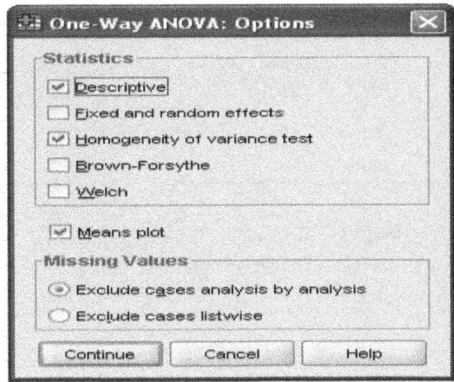

Figure 8.20 Screen showing option for descriptive statistics and testing assumption

Table 8.18   ANOVA Table for Testing Significance of Sex (Between-Subjects) Effects in Each Chocolate Category

Chocolate		Sum of Squares	df	Mean Square	F	Sig. (p-value)
Taste_Dark	Between groups	9.800	1	9.800	9.093	0.007
	Within groups	19.400	18	1.078		
	Total	29.200	19			
Taste_Milk	Between groups	33.800	1	33.800	32.021	0.000
	Within groups	19.000	18	1.056		
	Total	52.800	19			
Taste_White	Between groups	0.800	1	0.800	0.783	0.388
	Within groups	18.400	18	1.022		
	Total	19.200	19			

shown in Figure 8.20, click on **Continue** and then **OK** to get the results as shown in Tables 8.17 to 8.20.

Remark: The simple effect of Sex (between-subjects) can also be investigated by applying the *t* test for the independent groups in SPSS because sex has only two levels. But the one-way ANOVA approach has been adopted here so as to generalize the case.

Table 8.17 shows that the homogeneity of variance assumption has not been violated; hence ANOVA for independent measures can be applied for testing the effect of sex on Taste in each chocolate category separately.

Table 8.18 indicates that only the value of F for the Milk chocolate is significant as its associated p value is less than 0.0057. Hence, it can be concluded that the null hypothesis of no difference in mean scores on Taste between male and female groups is rejected in milk chocolate. Since F values for the Dark and White chocolates are not significant, it may be inferred that the male and female do not differ in their response on taste in the dark and the white chocolates.

Table 8.19   Descriptive Statistics

	N	Mean	Std. Dev.	Std. Error	95% Confidence Interval for Mean		Minimum	Maximum
					Lower Bound	Upper Bound		
Taste_Dark								
Male	10	5.1000	0.99443	0.31447	4.3886	5.8114	4.00	7.00
Female	10	6.5000	1.08012	0.34157	5.7273	7.2727	5.00	8.00
Total	20	5.8000	1.23969	0.27720	5.2198	6.3802	4.00	8.00
Taste_Milk								
Male	10	6.9000	0.87560	0.27689	6.2736	7.5264	5.00	8.00
Female	10	4.3000	1.15950	0.36667	3.4705	5.1295	3.00	6.00
Total	20	5.6000	1.66702	0.37276	4.8198	6.3802	3.00	8.00
Taste_White								
Male	10	5.6000	0.69921	0.22111	5.0998	6.1002	5.00	7.00
Female	10	6.0000	1.24722	0.39441	5.1078	6.8922	4.00	8.00
Total	20	5.8000	1.00525	0.22478	5.3295	6.2705	4.00	8.00

Table 8.20   Mauchly's Test of Sphericity in Different Sex

Measure: Crunchiness								
Sex	Within Subjects Effect	Mauchly's W	Approx. Chi-Square	df	Sig. (*p*-value)	Epsilon[a]		
						Greenhouse– Geisser	Huynh– Feldt	Lower-bound
Male	Chocolate	0.975	0.199	2	0.905	0.976	1.000	0.500
Female	Chocolate	0.895	0.891	2	0.641	0.905	1.000	0.500

Means Plots (Chocolate × Sex)   The means plot shown in the Figure 8.21 can be generated by using the outputs of the Tables 8.18 and 8.19. However, this plot is generated by the SPSS while solving the repeated MANOVA design in the illustration. This figure shows that males like the taste of milk chocolate more than females, whereas male and female do not find any difference in taste in dark chocolate as well as in white chocolate.

Simple Effect on Crunchiness   Analyzing simple effect on crunchiness will provide the correct picture about the subject's response in three different types of chocolates on its crunchiness in different sex groups. We shall investigate the simple effect of chocolate as well as sex on crunchiness. Simple effect of chocolate will facilitate us to compare response of the subjects on crunchiness among three different types of

ILLUSTRATION   227

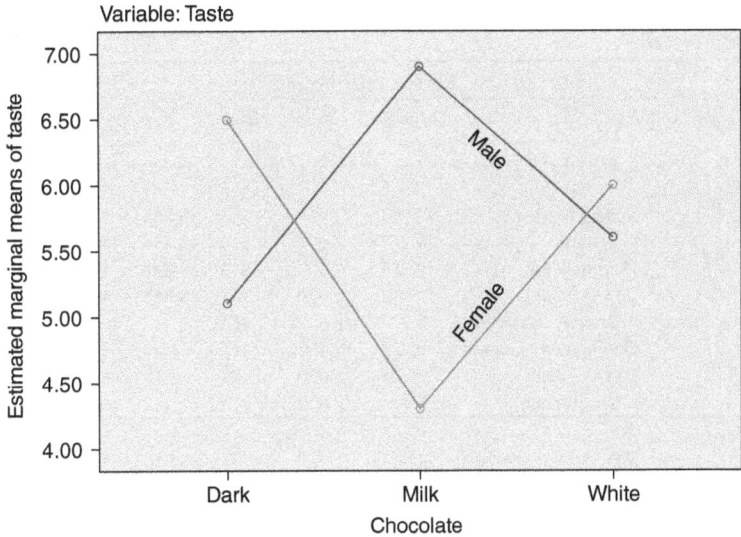

Figure 8.21   Marginal means plot of Chocolate×Sex for the data on Taste

chocolates in each sex group. Whereas, the simple effect of sex will provide us the comparison of the male and female response on the crunchiness in each type of the chocolate separately.

Effect of Chocolate (Within-Subjects)   The outputs for the simple effect of the choco-late (within-subjects) for the data on crunchiness can be obtained by following similar procedure we followed in finding the simple effect on Taste above. Table 8.20 indi-cates that the sphericity is not violated for the data in male and female category because none of the p values associated with Mauchly's W statistic is less than 0.017. Hence, we can state that the sphericity is assumed.

Table 8.21 indicates that the F values associated with the Chocolate (Sphericity assumed) in male ($p = 0.000$) and female ($p = 0.000$) sections are significant because p values associated with these F are less than 0.0085. Hence, pair-wise comparisons of marginal means shall be done in each sex category. Table 8.22 indicates the mean scores of crunchiness for each chocolate category in male and female groups, whereas Table 8.23 shows the pair-wise comparison of these marginal means. By combining the results of these tables marginal means plot of Sex × Chocolate for the data on Crunchiness as shown in the Figure 8.22 can be obtained. You need not generate this means plot because this plot is automatically generated in the outputs of the mixed design with two-way MANOVA in SPSS.

Means Plots (Sex × Chocolate)   We shall use the means plot shown in Figure 8.22 for comparing the mean scores of crunchiness in each type of chocolates in male and female groups separately. This means plot provides in-depth analysis of the response pattern. The plot indicates that the male subjects find milk chocolate to be signifi-cantly crunchier in comparison to that of white as well as dark chocolate. However,

Table 8.21   *F*-Table for Testing Significance of Chocolate (Within-Subjects) Effects in Each Sex Category

				Measure: Crunchiness			
Source Squared		Type III SS	df	Mean Square	F	Sig. (*p*-value)	Partial Eta Squared
Sex: Male							
Chocolate	Sphericity assumed	19.267	2	9.633	19.855	0.000	0.688
	Greenhouse–Geisser	19.267	1.952	9.870	19.855	0.000	0.688
	Huynh–Feldt	19.267	2.000	9.633	19.855	0.000	0.688
	Lower-bound	19.267	1.000	19.267	19.855	0.002	0.688
Error (Chocolate)	Sphericity assumed	8.733	18	0.485			
	Greenhouse–Geisser	8.733	17.569	0.497			
	Huynh–Feldt	8.733	18.000	0.485			
	Lower-bound	8.733	9.000	0.970			
Sex: Female							
Chocolate	Sphericity assumed	71.467	2	35.733	26.950	0.000	0.750
	Greenhouse–Geisser	71.467	1.809	39.498	26.950	0.000	0.750
	Huynh–Feldt	71.467	2.000	35.733	26.950	0.000	0.750
	Lower-bound	71.467	1.000	71.467	26.950	0.001	0.750
Error (Chocolate)	Sphericity assumed	23.867	18	1.326			
	Greenhouse–Geisser	23.867	16.284	1.466			
	Huynh–Feldt	23.867	18.000	1.326			
	Lower-bound	23.867	9.000	2.652			

Table 8.22   Descriptive Statistics

	Mean	Std. Deviation	N
Sex: Male			
Crunch_Dark	5.0000	0.66667	10
Crunch_Milk	6.7000	0.94868	10
Crunch_White	5.0000	0.66667	10
Sex: Female			
Crunch_Dark	7.8000	1.03280	10
Crunch_Milk	5.0000	1.49071	10
Crunch_White	4.2000	0.78881	10

the trend is reversed in the female section. Female subjects find dark chocolate to be the crunchiest among all the three types of chocolates. However, there is no difference between the women's response on crunchiness of the milk and white chocolates.

Effect of Sex (Between-Subjects)   The outputs for the simple effect of the Sex (between-subjects) for the data on crunchiness can be obtained by following the same procedure we did in case of generating the simple effect of Sex (between-subjects) on Taste above. Table 8.24 shows that the homogeneity of variance assumption is not violated because none of the *p* values associated with the Levene's statistic is less than 0.05. Thus, ANOVA for independent measures can be applied for comparing

ILLUSTRATION 229

Table 8.23   Pair-wise Comparisons of Marginal Means in Each Sex Category

Measure: Crunchiness						
		Mean Diff.			95% CI for Difference[a]	
Chocolate (I)	Chocolate (J)	(I − J)	Std. Error	Sig[a]	Lower Bound	Upper Bound
Sex: Male						
Dark	Milk	−1.700*	0.335	0.002	−2.683	−0.717
	White	0.000	0.298	1.000	−0.875	0.875
Milk	Dark	1.700*	0.335	0.002	0.717	2.683
	White	1.700*	0.300	0.001	0.820	2.580
White	Dark	0.000	0.298	1.000	−0.875	0.875
	Milk	−1.700*	0.300	0.001	−2.580	−0.820
Sex: Female						
Dark	Milk	2.800*	0.467	0.001	1.431	4.169
	White	3.600*	0.476	0.000	2.203	4.997
Milk	Dark	−2.800*	0.467	0.001	−4.169	−1.431
	White	0.800	0.593	0.630	−0.938	2.538
White	Dark	−3.600*	0.476	0.000	−4.997	−2.203
	Milk	−0.800	0.593	0.630	−2.538	0.938

Based on estimated marginal means.
*The mean difference is significant at the 0.05 level.
[a]Adjustment for multiple comparisons: Bonferroni.

Table 8.24   Test of Homogeneity of Variances

	Levene Statistic	df1	df2	Sig. (p-value)
Crunch_Dark	3.524	1	18	0.077
Crunch_Milk	2.199	1	18	0.155
Crunch_White	1.328	1	18	0.264

the response of male and female on crunchiness in each of the three categories of the chocolate.

Table 8.25 shows that the F value for the Dark ($p = 0.000$) chocolate is significant as the p value associated with it is less than 0.0057. Thus, it can be concluded that the null hypothesis of no difference in mean scores on Crunchiness between male and female groups is rejected at 5% level in dark chocolate. However, F values for the Milk and White chocolates are not significant; hence it may be inferred that the male and female do not differ in their response on crunchiness in the milk and white chocolates.

Means Plots (Chocolate × Sex)   The means plot shown in the Figure 8.23 can be generated by using the results of the Tables 8.25 and 8.26. This plot indicates that the female response on the crunchiness of the dark chocolate is significantly higher than

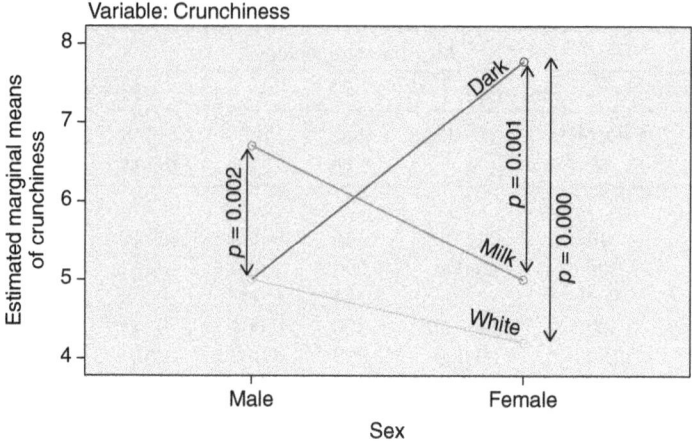

Figure 8.22   Marginal means plot of Sex × Chocolate for the data on Crunchiness

Table 8.25   *F*-table for Testing Significance of Sex (Between-Subjects) Effect
on Crunchiness in Each Chocolate Category

Chocolate		Sum of Squares	df	Mean Square	F	Sig. (*p*-value)
Crunch_Dark	Between groups	39.200	1	39.200	51.882	0.000
	Within groups	13.600	18	0.756		
	Total	52.800	19			
Crunch_Milk	Between groups	14.450	1	14.450	9.256	0.007
	Within groups	28.100	18	1.561		
	Total	42.550	19			
Crunch_White	Between groups	3.200	1	3.200	6.000	0.025
	Within groups	9.600	18	0.533		
	Total	12.800	19			

that of the male. However, the response of male and female does not differ in the milk as well as in the white chocolates.

Simple Effect on Flavor   Simple effect on flavor will provide the detailed picture about the subject's response in three different types of chocolates on its flavor in different sex groups. We shall investigate the simple effect of chocolate and sex on flavor. Simple effect of chocolate will facilitate us to compare the subject's response on the flavor among three different types of chocolates in each sex group. Whereas, the simple effect of sex will facilitate us to compare the response of male and female on the flavor in each type of the chocolate separately.

Effect of Chocolate (Within-Subjects)   Table 8.27 indicates that the sphericity is not violated for the data in male and female category; hence we can state that the sphericity is assumed.

ILLUSTRATION   **231**

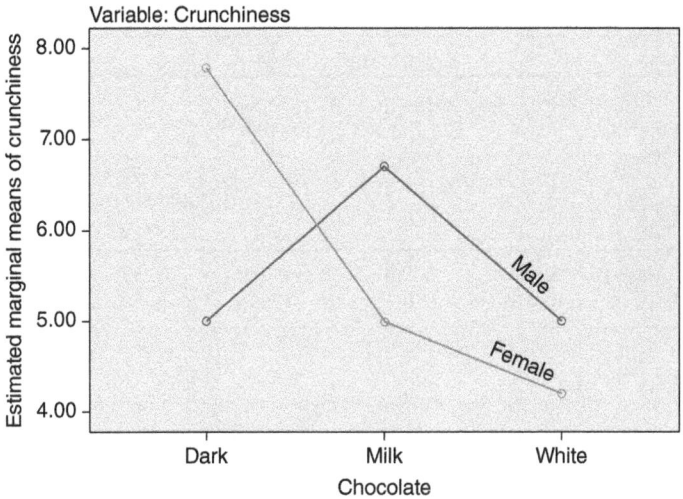

Figure 8.23   Marginal means plot of Chocolate × Sex for the data on Crunchiness

Table 8.26   Descriptive Statistics

	N	Mean	Std. Dev.	Std. Error	95% Confidence Interval for Mean		Minimum	Maximum
					Lower Bound	Upper Bound		
Crunch_Dark								
Male	10	5.0000	0.66667	0.21082	4.5231	5.4769	4.00	6.00
Female	10	7.8000	1.03280	0.32660	7.0612	8.5388	6.00	9.00
Total	20	6.4000	1.66702	0.37276	5.6198	7.1802	4.00	9.00
Crunch_Milk								
Male	10	6.7000	0.94868	0.30000	6.0214	7.3786	5.00	8.00
Female	10	5.0000	1.49071	0.47140	3.9336	6.0664	3.00	8.00
Total	20	5.8500	1.49649	0.33462	5.1496	6.5504	3.00	8.00
Crunch_White								
Male	10	5.0000	0.66667	0.21082	4.5231	5.4769	4.00	6.00
Female	10	4.2000	0.78881	0.24944	3.6357	4.7643	3.00	5.00
Total	20	4.6000	0.82078	0.18353	4.2159	4.9841	3.00	6.00

Table 8.28 reveals that the F values associated with the Chocolate (Sphericity assumed) in male ($p = 0.000$) and female ($p = 0.001$) sections are significant because p-values associated with these F are less than 0.0085. Now the pair-wise comparison of marginal means shall be done in each sex category. Table 8.29 indicates the mean scores of flavor for each chocolate category in male and female groups, whereas Table 8.30 shows the pair-wise comparison of these marginal means. By combining the results of these tables the marginal means plot of Sex × Chocolate for the data

Table 8.27 Mauchly's Test of Sphericity in Different Sex

Sex	Within Subjects Effect	Mauchly's W	Approx. Chi-Square	df	Sig. (p-value)	Epsilon		
						Greenhouse–Geisser	Huynh–Feldt	Lower-bound
Male	Chocolate	0.968	0.256	2	0.880	0.969	1.000	0.500
Female	Chocolate	0.956	0.361	2	0.835	0.958	1.000	0.500

Measure: Flavor

on Flavor as shown in the Figure 8.24 can be obtained. This means plot in fact is generated in the main outputs of the analysis in SPSS.

Means Plots (Sex × Chocolate)   The means plot shown in Figure 8.24 shall be used for comparing the mean scores of flavor in each type of chocolate in male and female groups separately. This figure indicates that the male subjects find milk chocolate to be more flavored in comparison to that of white as well as dark chocolate. However, the trend is reversed in the female section. Female subjects find dark chocolate to be more flavored in comparison to milk chocolate.

Table 8.28 *F*-Table for Testing Significance of Chocolate (Within-Subjects) Effects in Each Sex Category

Measure: Flavor

Source		Type III SS	df	Mean Square	F	Sig. (p-value)	Partial Eta Squared
Sex: Male							
Chocolate	Sphericity assumed	29.600	2	14.800	22.705	0.000	0.716
	Greenhouse–Geisser	29.600	1.939	15.266	22.705	0.000	0.716
	Huynh–Feldt	29.600	2.000	14.800	22.705	0.000	0.716
	Lower-bound	29.600	1.000	29.600	22.705	0.001	0.716
Error (Chocolate)	Sphericity assumed	11.733	18	0.652			
	Greenhouse–Geisser	11.733	17.450	0.672			
	Huynh–Feldt	11.733	18.000	0.652			
	Lower-bound	11.733	9.000	1.304			
Sex: Female							
Chocolate	Sphericity assumed	8.267	2	4.133	10.528	0.001	0.539
	Greenhouse–Geisser	8.267	1.915	4.316	10.528	0.001	0.539
	Huynh–Feldt	8.267	2.000	4.133	10.528	0.001	0.539
	Lower-bound	8.267	1.000	8.267	10.528	0.010	0.539
Error (Chocolate)	Sphericity assumed	7.067	18	0.393			
	Greenhouse–Geisser	7.067	17.239	0.410			
	Huynh–Feldt	7.067	18.000	0.393			
	Lower-bound	7.067	9.000	0.785			

ILLUSTRATION   233

Table 8.29 Descriptive statistics

	Mean	Std. Deviation	N
Sex: Male			
Flavor_Dark	5.3000	1.05935	10
Flavor_Milk	7.3000	0.67495	10
Flavor_White	5.1000	0.73786	10
Sex: Female			
Flavor_Dark	6.3000	1.05935	10
Flavor_Milk	5.1000	0.73786	10
Flavor_White	5.3000	0.67495	10

Simple Effect of Sex (Between-Subjects) Table 8.31 reveals that the homogeneity of variance assumption is not been violated because none of the p values associated with the Levene's statistic is less than 0.05. Thus, ANOVA for independent measures can be applied for comparing the response of male and female on flavor in each of the three categories of chocolate.

Table 8.32 shows that the F value for the Milk chocolate ($p = 0.000$) only is significant as the p value associated with it is less than 0.0057. Thus, it may be concluded that the null hypothesis of no difference in mean scores on the Flavor between male

Table 8.30 Pair-wise Comparisons of Marginal Means in Each Sex Category

			Measure: Flavor			
		Mean Diff.			95% CI for Difference[a]	
Chocolate (I)	Chocolate (J)	(I − J)	Std. Error	Sig[a]	Lower Bound	Upper Bound
Sex: Male						
Dark	Milk	−2.000*	0.333	0.001	−2.978	−1.022
	White	0.200	0.389	1.000	−0.940	1.340
Milk	Dark	2.000*	0.333	0.001	1.022	2.978
	White	2.200*	0.359	0.001	1.147	3.253
White	Dark	−0.200	0.389	1.000	−1.340	0.940
	Milk	−2.200*	0.359	0.001	−3.253	−1.147
Sex: Female						
Dark	Milk	1.200*	0.249	0.003	0.468	1.932
	White	1.000*	0.298	0.025	0.125	1.875
Milk	Dark	−1.200*	0.249	0.003	−1.932	−0.468
	White	−0.200	0.291	1.000	−1.052	0.652
White	Dark	−1.000*	0.298	0.025	−1.875	−0.125
	Milk	0.200	0.291	1.000	−0.652	1.052

Based on estimated marginal means.
*The mean difference is significant at the 0.0085 level.
[a]Adjustment for multiple comparisons: Bonferroni.

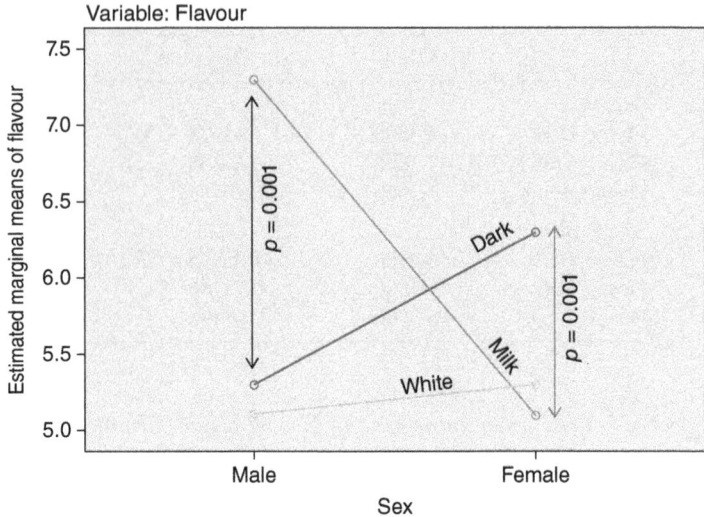

Figure 8.24   Marginal means plot of Sex × Chocolate for the data on Flavor

and female groups is rejected in relation to milk chocolate. However, F values for the dark and white chocolates are not significant.

Means Plots (Chocolate × Sex) The means plot shown in Figure 8.25 may be used for in-depth analysis. It can be generated by using the results of the Tables 8.32 and 8.33. This plot indicates that the male response on the flavor of the milk chocolate is significantly higher than that of the female.

How to Report Findings

In this illustration, mixed design with two-way MANOVA was used to investigate the effects of Chocolate (within-subjects), Sex (between-subjects), and the interaction between Chocolate and Sex on a group of dependent variables (taste, crunchiness, and flavor). These hypotheses were tested at the family-wise error rate 0.05.

Assumptions The assumptions of the mixed design with two-way MANOVA were tested which are reported as follows:

Table 8.31   Test of Homogeneity of Variances

	Levene Statistic	df1	df2	Sig. (p-value)
Flavor_Dark	0.107	1	18	0.747
Flavor_Milk	0.012	1	18	0.913
Flavor_White	0.012	1	18	0.913

ILLUSTRATION   235

Table 8.32 *F*-Table for Testing Significance of Sex (Between-Subjects) Effects in Each Chocolate Category

Chocolate		Sum of Squares	df	Mean Square	F	Sig. (*p*-value)
Flavor_Dark	Between groups	5.000	1	5.000	4.455	0.049
	Within groups	20.200	18	1.122		
	Total	25.200	19			
Flavor_Milk	Between groups	24.200	1	24.200	48.400	0.000
	Within groups	9.000	18	0.500		
	Total	33.200	19			
Flavor_White	Between groups	0.200	1	0.200	0.400	0.535
	Within groups	9.000	18	0.500		
	Total	9.200	19			

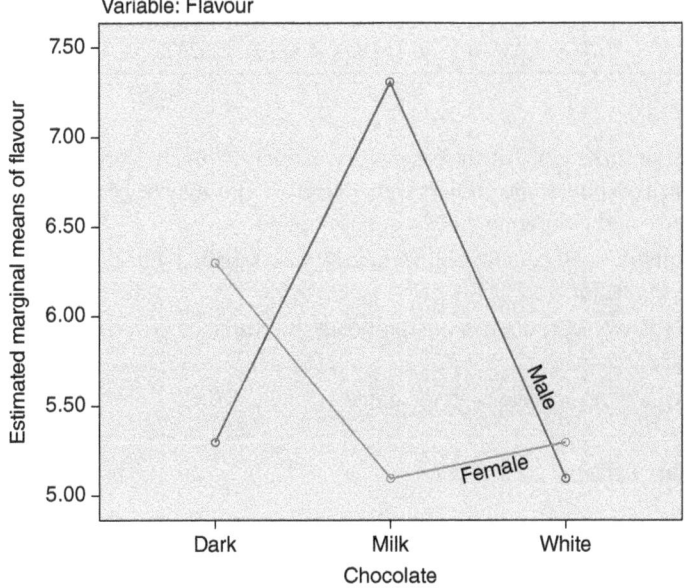

Figure 8.25 Marginal means plot of Chocolate × Sex for the data on Flavor

- Since the independent variables, Chocolate and Sex, were categorical and dependent variables (response on the taste, crunchiness, and flavor) were continuous, the assumption about the data type holds true.
- A mixed trend of relationships was observed among the dependent variables. This may be due to the small number of the subjects.
- Since Shapiro–Wilk statistic was found to be significant for the data set of the variables Crunch_Dark, Taste_Milk, Flavor_Milk, Taste_White, Crunch_White, and Flavor_White, data in these cell was not normally distributed.

Table 8.33   Descriptive Statistics

N		Mean	Std. Dev.	Std. Error	95% Confidence Interval for Mean		Minimum	Maximum
					Lower Bound	Upper Bound		
Flavor_Dark								
Male	10	5.3000	1.05935	0.33500	4.5422	6.0578	4.00	7.00
Female	10	6.3000	1.05935	0.33500	5.5422	7.0578	5.00	8.00
Total	20	5.8000	1.15166	0.25752	5.2610	6.3390	4.00	8.00
Flavor_Milk								
Male	10	7.3000	0.67495	0.21344	6.8172	7.7828	6.00	8.00
Female	10	5.1000	0.73786	0.23333	4.5722	5.6278	4.00	6.00
Total	20	6.2000	1.32188	0.29558	5.5813	6.8187	4.00	8.00
Flavor_White								
Male	10	5.1000	0.73786	0.23333	4.5722	5.6278	4.00	6.00
Female	10	5.3000	0.67495	0.21344	4.8172	5.7828	4.00	6.00
Total	20	5.2000	0.69585	0.15560	4.8743	5.5257	4.00	6.00

- Only in three cells of male category, namely Crunch_Dark, Taste_Milk, and Crunch_White, some outliers were detected. This may be because of small sample size and narrow range of response option.
- Assumption of equality of variances was satisfied for comparing between-subjects factor.
- Since Box's M test was not significant, the variance covariance matrices were equal.
- There was no sphericity in the data.

Multivariate Effects

- Multivariate effect of between-subjects factor (sex) on the combined characteristics of chocolate was not significant irrespective of the chocolate group: Wilks' $\lambda = 0.898$, $F(3,16) = 0.603$, $p = 0.623$
- There was a significant multivariate effect of within-subjects factor (chocolate) on the combined characteristics of chocolate irrespective of the sex: Wilks' $\lambda = 0.206$, $F(6, 13) = 8.369$, $p = 0.001$.
- There was also a significant multivariate effect across the interaction between the sex group and chocolate group: Wilks' $\lambda = 0.060$, $F(6, 13) = 34.203$, $p = 0.000$.

Univariate Main Effects   Since sphericity assumption was not violated for the data on all the three dependent variables for comparing the performance across the chocolate groups, no correction was applied.

ILLUSTRATION 237

- The effect of Sex (between-subjects) was not significant on any of the chocolate characteristics, that is, Taste, Crunchiness, and Flavor irrespective of the chocolate types (dark, milk, and white)
- The effects of Chocolate on Crunchiness ($p = 0.000$) and Flavor ($p = 0.000$) were significant irrespective of the sexes; however, no significant effect of Chocolate on taste was observed.
- The interaction effect (chocolate × sex) was significant in each of the three dependent variables; Taste ($p = 0.000$), Crunchiness ($p = 0.000$), and Flavor ($p = 0.000$)

Univariate Simple Effects
Simple effect on Taste

- Response of male subjects on the taste of the milk chocolate was significantly higher than that of white and dark chocolates.
- Response of the female subjects on the taste of the dark chocolate was significantly higher than that of the milk chocolate.
- Males' response on the taste of the milk chocolate was significantly higher than that of the females.

Simple effect on Crunchiness

- Response of male subjects on the crunchiness of the milk chocolate was significantly higher than that of white and dark chocolates.
- Response of the female subjects on the crunchiness of the dark chocolate was significantly higher than that of the milk and white chocolates.
- Females' response on the crunchiness of the dark chocolate was significantly higher than that of the males.

Simple effect on Flavor

- Response of male subjects on the flavor of the milk chocolate was significantly higher than that of the dark and white chocolates.
- Response of the female subjects on the flavor of the dark chocolate was significantly higher than that of the white as well as milk chocolates.
- Males' response on the flavor of the milk chocolate was significantly higher than that of the females.

Inference

On the basis of the sample observation it may be inferred that the response on the chocolate characteristics varied among three different types of chocolates, that is, dark, white, and milk. Response pattern in three different types of chocolates on the overall chocolate characteristics varied in male and female.

There was no gender preference on the response of each of the chocolate characteristics irrespective of the type of the chocolates. However, response of the subjects on the crunchiness and flavor among the three types of chocolates differs irrespective of the sexes. The pattern of response on each chocolate characteristic in three different types of chocolates was different in male and female. The males' response on the taste of the milk chocolate was significantly higher than that of the females, whereas no such gender preferences were observed on taste in dark and white chocolates.

Male subjects found taste of the milk chocolate to be the best, whereas female subjects found dark chocolate to be the tastiest. Male subjects found milk chocolate to be the crunchiest, whereas female responded the other way round and found dark chocolate to be the crunchiest.

In the opinion of males the flavor of milk chocolate was the best, whereas females preferred flavor of the dark chocolates. The males' response on the flavor of the milk chocolate was significantly higher than that of the females, whereas no such preference was observed in the response for the dark and white chocolates.

EXERCISE

8.1. Explain situations where repeated measures MANOVA can be used.

8.2. An experimenter organized a mixed MANOVA design to investigate the effect of gender and brand on pizza quality. Six male and six female subjects were selected in a study. Each subject was asked to eat all the three pizza and rate their quality on three parameters; taste, spicy, and flavor. Show the layout design in the study. Frame the research questions to be investigated.

8.3. Discuss briefly the steps involved in solving a mixed MANOVA design. What all outputs you would like to generate if SPSS is used for solving this design?

8.4. Identify a situation where mixed MANOVA design can be used to study the pattern of response by subjects over the period of time and construct the hypotheses to be tested in the design.

8.5. How the issue of inflating family-wise error rate is taken care of in this design while doing the univariate analysis and investigating the simple effect?

ASSIGNMENT

8.1. Consider a study in which the effectiveness of a new antidepressant drug was tested for two months on the patients suffering from depression. Twelve patients were randomly divided into two groups and the new drug was administered on the six patients in the first group, whereas second group got

Table 8.34 Response on the symptoms of depression

Treatment	Subject	Initial			One month			Two month		
		hopelessness	irritability	cognitions	hopelessness	irritability	cognitions	hopelessness	irritability	cognitions
Antidepressant drug	1	16	16	15	15	12	13	10	11	11
	2	15	14	14	14	11	14	12	12	13
	3	14	15	16	14	12	15	12	11	10
	4	15	16	17	13	12	13	11	10	9
	5	15	17	16	15	16	14	10	9	10
	6	14	16	14	13	15	14	12	11	11
Placebo	1	17	16	17	14	15	16	17	15	15
	2	16	18	16	13	16	15	15	16	14
	3	18	17	16	13	16	15	18	15	15
	4	16	18	18	15	17	16	17	16	17
	5	15	19	16	14	18	15	15	15	16
	6	17	17	15	16	16	14	16	14	15

the placebo. Each subject in both the groups was tested for the symptoms of depression (hopelessness, irritability, and cognitions) before the experiment, after one, and two months of administering the treatments. The data so obtained in the experiment is shown in the Table 8.34. Apply two-factor mixed MANOVA to answer the following research questions by testing the appropriate hypotheses at 5% level.

a. Whether new drug is effective in reducing the depression in patients?

b. Whether new drug is effective in reducing each depression parameters?

c. Whether interaction between the drug and time is significant?

d. Which duration is more appropriate in improving depression parameters?

BIBLIOGRAPHY

Bird KD, Hadzi-Pavlovic D. Simultaneous test procedures and the choice of a test statistic in MANOVA. Psychol Bull 1983;93:167–178.

Bock RD. *Multivariate statistical methods in behavioral research.* New York: McGraw-Hill; 1975.

Boik RJ. A priori tests in repeated measures designs: Effects of nonsphericity. Psychometrika 1981;46:241–255.

Collier RO, Baker EB, MandeviUe GK, Hayes TE. Estimates of test size for several procedures based on conventional ratios in the repeated measure design. Psychometrika 1967;32:339–353.

Davidson ML. Univariate versus multivariate tests in repeated measures experiments. Psychol Bull 1972;77:446–452.

Harris RJ. *A primer of multivariate statistics.* New York: Academic Press; 1975.

Huynh H, Mandeville GK. Validity conditions in repeated measures designs. Psychol Bull 1979;86:964–973.

Mardia KV, Kent JT, Bibby JM. *Multivariate Analysis.* Academic Press ISBN 0-12-471252-5; 1979.

Keselman HJ, Rogan JC, Mendoza JL, Breen LL. Testing the validity conditions of repeated measures F tests. Psychol Bull 1980;87:479–481.

Leech NL, Barrett KC, Morgan GA. *SPSS for Intermediate Statistics: Use and Interpretation.* 2nd ed. Mahwah, NJ: Lawrence Erlbaum Associates; 2005.

Maxwell SE. Pairwise multiple comparisons in repeated measures designs. J Educ Stat 1980;5:269–287.

McCall RB, Appelbaum MI. Bias in the analysis of repeated-measures designs: some alternative approaches. Child Dev 1973;44:401–415.

Mitzel HC, Games PA. Circularity and multiple comparisons in repeated measures designs. Br J Math Stat Psych 1981;34:253–259.

Olson CL. Comparative robustness of six tests in multivariate analysis of variance. J Am Stat Assoc 1974;69:894–908.

Rngan JC, Keselman HJ, Mendoza JL. Analysis of repeated measurements. Br J Math Stat Psych 1979;32:269–286.

Roy SN. *Some Aspects of Multivariate Analysis.* New York: Wiley; 1957.

Timm NH. Multivariate analysis of variance of repeated measurements. In: Krishnaiah ER, editor. *Handbook of Statistics. Volume I: Analysis of Variance.* New York: Elsevier-North Holland; 1980. p 41–87.

Wallenstein S, Fleiss JL. Repeated measures analysis of variance when the correlations have a certain pattern. Psychometrika 1979;44:229–233.

APPENDIX

See Tables A.1–A.4

Table A.1 The Normal Curve Area between the Mean and a Given z Value

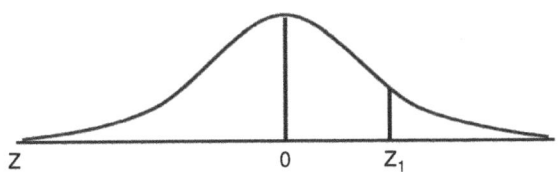

Z	0.00	0.01	0.02	0.03	0.04	0.05	0.06	0.07	0.08	0.09
0.0	0.0000	0.0040	0.0080	0.0120	0.0160	0.0199	0.0239	0.0279	0.0319	0.0359
0.1	0.0398	0.0438	0.0478	0.0517	0.0557	0.0596	0.0636	0.0675	0.0714	0.0753
0.2	0.0793	0.0832	0.0871	0.0910	0.0948	0.0987	0.1026	0.1064	0.1103	0.1141
0.3	0.1179	0.1217	0.1255	0.1293	0.1331	0.1368	0.1406	0.1443	0.1480	0.1517
0.4	0.1554	0.1591	0.1628	0.1664	0.1700	0.1736	0.1772	0.1808	0.1844	0.1879
0.5	0.1915	0.1950	0.1985	0.2019	0.2054	0.2088	0.2123	0.2157	0.2190	0.2224
0.6	0.2257	0.2291	0.2324	0.2357	0.2389	0.2422	0.2454	0.2486	0.2517	0.2549

(continued)

Repeated Measures Design for Empirical Researchers, First Edition. J. P. Verma.
© 2016 John Wiley & Sons, Inc. Published 2016 by John Wiley & Sons, Inc.

Table A.1 (*Continued*)

Z	0.00	0.01	0.02	0.03	0.04	0.05	0.06	0.07	0.08	0.09
0.7	0.2580	0.2611	0.2642	0.2673	0.2704	0.2734	0.2764	0.2794	0.2823	0.2852
0.8	0.2881	0.2910	0.2939	0.2967	0.2995	0.3023	0.3051	0.3078	0.3106	0.3133
0.9	0.3159	0.3186	0.3212	0.3238	0.3264	0.3289	0.3315	0.3340	0.3365	0.3389
1.0	0.3413	0.3438	0.3461	0.3485	0.3508	0.3531	0.3554	0.3577	0.3599	0.3621
1.1	0.3643	0.3665	0.3686	0.3708	0.3729	0.3749	0.3770	0.3790	0.3810	0.3830
1.2	0.3849	0.3869	0.3888	0.3907	0.3925	0.3944	0.3962	0.3980	0.3997	0.4015
1.3	0.4032	0.4049	0.4066	0.4082	0.4099	0.4115	0.4131	0.4147	0.4162	0.4177
1.4	0.4192	0.4207	0.4222	0.4236	0.4251	0.4265	0.4279	0.4292	0.4306	0.4319
1.5	0.4332	0.4345	0.4357	0.4370	0.4382	0.4394	0.4406	0.4418	0.4429	0.4441
1.6	0.4452	0.4463	0.4474	0.4484	0.4495	0.4505	0.4515	0.4525	0.4535	0.4545
1.7	0.4554	0.4564	0.4573	0.4582	0.4591	0.4599	0.4608	0.4616	0.4625	0.4633
1.8	0.4641	0.4649	0.4656	0.4664	0.4671	0.4678	0.4686	0.4693	0.4699	0.4706
1.9	0.4713	0.4719	0.4726	0.4732	0.4738	0.4744	0.4750	0.4756	0.4761	0.4767
2.0	0.4772	0.4778	0.4783	0.4788	0.4793	0.4798	0.4803	0.4808	0.4812	0.4817
2.1	0.4821	0.4826	0.4830	0.4834	0.4838	0.4842	0.4846	0.4850	0.4854	0.4857
2.2	0.4861	0.4864	0.4868	0.4871	0.4875	0.4878	0.4881	0.4884	0.4887	0.4890
2.3	0.4893	0.4896	0.4898	0.4901	0.4904	0.4906	0.4909	0.4911	0.4913	0.4916
2.4	0.4918	0.4920	0.4922	0.4925	0.4927	0.4929	0.4931	0.4932	0.4934	0.4936
2.5	0.4938	0.4940	0.4941	0.4943	0.4945	0.4946	0.4948	0.4949	0.4951	0.4952
2.6	0.4953	0.4955	0.4956	0.4957	0.4959	0.4960	0.4961	0.4962	0.4963	0.4964
2.7	0.4965	0.4966	0.4967	0.4968	0.4969	0.4970	0.4971	0.4972	0.4973	0.4974
2.8	0.4974	0.4975	0.4976	0.4977	0.4977	0.4978	0.4979	0.4979	0.4980	0.4981
2.9	0.4981	0.4982	0.4982	0.4983	0.4984	0.4984	0.4985	0.4985	0.4986	0.4986
3.0	0.4987	0.4987	0.4987	0.4988	0.4988	0.4989	0.4989	0.4989	0.4990	0.4990

Table A.2   *F*-Table: Critical Values $\alpha = 0.05$

v_1/v_2	1	2	3	4	5	6	7	8	9	10	11	12
3	10.13	9.55	9.28	9.12	9.01	8.94	8.89	8.85	8.81	8.79	8.76	8.74
4	7.71	6.94	6.59	6.39	6.26	6.16	6.09	6.04	6.00	5.96	5.94	5.91
5	6.61	5.79	5.41	5.19	5.05	4.95	4.88	4.82	4.77	4.74	4.70	4.68
6	5.99	5.14	4.76	4.53	4.39	4.28	4.21	4.15	4.10	4.06	4.03	4.00
7	5.59	4.74	4.35	4.12	3.97	3.87	3.79	3.73	3.68	3.64	3.60	3.57
8	5.32	4.46	4.07	3.84	3.69	3.58	3.50	3.44	3.39	3.35	3.31	3.28
9	5.12	4.26	3.86	3.63	3.48	3.37	3.29	3.23	3.18	3.14	3.10	3.07
10	4.96	4.10	3.71	3.48	3.33	3.22	3.14	3.07	3.02	2.98	2.94	2.91
11	4.84	3.98	3.59	3.36	3.20	3.09	3.01	2.95	2.90	2.85	2.82	2.79
12	4.75	3.89	3.49	3.26	3.11	3.00	2.91	2.85	2.80	2.75	2.72	2.69
13	4.67	3.81	3.41	3.18	3.03	2.92	2.83	2.77	2.71	2.67	2.63	2.60
14	4.60	3.74	3.34	3.11	2.96	2.85	2.76	2.70	2.65	2.60	2.57	2.53
15	4.54	3.68	3.29	3.06	2.90	2.79	2.71	2.64	2.59	2.54	2.51	2.48
16	4.49	3.63	3.24	3.01	2.85	2.74	2.66	2.59	2.54	2.49	2.46	2.42
17	4.45	3.59	3.20	2.96	2.81	2.70	2.61	2.55	2.49	2.45	2.41	2.38
18	4.41	3.55	3.16	2.93	2.77	2.66	2.58	2.51	2.46	2.41	2.37	2.34
19	4.38	3.52	3.13	2.90	2.74	2.63	2.54	2.48	2.42	2.38	2.34	2.31
20	4.35	3.49	3.10	2.87	2.71	2.60	2.51	2.45	2.39	2.35	2.31	2.28
22	4.30	3.44	3.05	2.82	2.66	2.55	2.46	2.40	2.34	2.30	2.26	2.23
24	4.26	3.40	3.01	2.78	2.62	2.51	2.42	2.36	2.30	2.25	2.22	2.18
26	4.23	3.37	2.98	2.74	2.59	2.47	2.39	2.32	2.27	2.22	2.18	2.15
28	4.20	3.34	2.95	2.71	2.56	2.45	2.36	2.29	2.24	2.19	2.15	2.12
30	4.17	3.32	2.92	2.69	2.53	2.42	2.33	2.27	2.21	2.16	2.13	2.09
35	4.12	3.27	2.87	2.64	2.49	2.37	2.29	2.22	2.16	2.11	2.08	2.04
40	4.08	3.23	2.84	2.61	2.45	2.34	2.25	2.18	2.12	2.08	2.04	2.00
45	4.06	3.20	2.81	2.58	2.42	2.31	2.22	2.15	2.10	2.05	2.01	1.97
50	4.03	3.18	2.79	2.56	2.40	2.29	2.20	2.13	2.07	2.03	1.99	1.95
60	4.00	3.15	2.76	2.53	2.37	2.25	2.17	2.10	2.04	1.99	1.95	1.92
70	3.98	3.13	2.74	2.50	2.35	2.23	2.14	2.07	2.02	1.97	1.93	1.89
80	3.96	3.11	2.72	2.49	2.33	2.21	2.13	2.06	2.00	1.95	1.91	1.88
100	3.94	3.09	2.70	2.46	2.31	2.19	2.10	2.03	1.97	1.93	1.89	1.85
200	3.89	3.04	2.65	2.42	2.26	2.14	2.06	1.98	1.93	1.88	1.84	1.80
500	3.86	3.01	2.62	2.39	2.23	2.12	2.03	1.96	1.90	1.85	1.81	1.77
1000	3.85	3.00	2.61	2.38	2.22	2.11	2.02	1.95	1.89	1.84	1.80	1.76
>1000	1.04	3.00	2.61	2.37	2.21	2.10	2.01	1.94	1.88	1.83	1.79	1.75

(continued)

Table A.2 (*Continued*)

v_1/v_2	13	14	15	16	17	18	19	20	22	24	26	28
3	8.73	8.71	8.70	8.69	8.68	8.67	8.67	8.66	8.65	8.64	8.63	8.62
4	5.89	5.87	5.86	5.84	5.83	5.82	5.81	5.80	5.79	5.77	5.76	5.75
5	4.66	4.64	4.62	4.60	4.59	4.58	4.57	4.56	4.54	4.53	4.52	4.50
6	3.98	3.96	3.94	3.92	3.91	3.90	3.88	3.87	3.86	3.84	3.83	3.82
7	3.55	3.53	3.51	3.49	3.48	3.47	3.46	3.44	3.43	3.41	3.40	3.39
8	3.26	3.24	3.22	3.20	3.19	3.17	3.16	3.15	3.13	3.12	3.10	3.09
9	3.05	3.03	3.01	2.99	2.97	2.96	2.95	2.94	2.92	2.90	2.89	2.87
10	2.89	2.86	2.85	2.83	2.81	2.80	2.79	2.77	2.75	2.74	2.72	2.71
11	2.76	2.74	2.72	2.70	2.69	2.67	2.66	2.65	2.63	2.61	2.59	2.58
12	2.66	2.64	2.62	2.60	2.58	2.57	2.56	2.54	2.52	2.51	2.49	2.48
13	2.58	2.55	2.53	2.51	2.50	2.48	2.47	2.46	2.44	2.42	2.41	2.39
14	2.51	2.48	2.46	2.44	2.43	2.41	2.40	2.39	2.37	2.35	2.33	2.32
15	2.45	2.42	2.40	2.38	2.37	2.35	2.34	2.33	2.31	2.29	2.27	2.26
16	2.40	2.37	2.35	2.33	2.32	2.30	2.29	2.28	2.25	2.24	2.22	2.21
17	2.35	2.33	2.31	2.29	2.27	2.26	2.24	2.23	2.21	2.19	2.17	2.16
18	2.31	2.29	2.27	2.25	2.23	2.22	2.20	2.19	2.17	2.15	2.13	2.12
19	2.28	2.26	2.23	2.21	2.20	2.18	2.17	2.16	2.13	2.11	2.10	2.08
20	2.25	2.23	2.20	2.18	2.17	2.15	2.14	2.12	2.10	2.08	2.07	2.05
22	2.20	2.17	2.15	2.13	2.11	2.10	2.08	2.07	2.05	2.03	2.01	2.00
24	2.15	2.13	2.11	2.09	2.07	2.05	2.04	2.03	2.00	1.98	1.97	1.95
26	2.12	2.09	2.07	2.05	2.03	2.02	2.00	1.99	1.97	1.95	1.93	1.91
28	2.09	2.06	2.04	2.02	2.00	1.99	1.97	1.96	1.93	1.91	1.90	1.88
30	2.06	2.04	2.01	1.99	1.98	1.96	1.95	1.93	1.91	1.89	1.87	1.85
35	2.01	1.99	1.96	1.94	1.92	1.91	1.89	1.88	1.85	1.83	1.82	1.80
40	1.97	1.95	1.92	1.90	1.89	1.87	1.85	1.84	1.81	1.79	1.77	1.76
45	1.94	1.92	1.89	1.87	1.86	1.84	1.82	1.81	1.78	1.76	1.74	1.73
50	1.92	1.89	1.87	1.85	1.83	1.81	1.80	1.78	1.76	1.74	1.72	1.70
60	1.89	1.86	1.84	1.82	1.80	1.78	1.76	1.75	1.72	1.70	1.68	1.66
70	1.86	1.84	1.81	1.79	1.77	1.75	1.74	1.72	1.70	1.67	1.65	1.64
80	1.84	1.82	1.79	1.77	1.75	1.73	1.72	1.70	1.68	1.65	1.63	1.62
100	1.82	1.79	1.77	1.75	1.73	1.71	1.69	1.68	1.65	1.63	1.61	1.59
200	1.77	1.74	1.72	1.69	1.67	1.66	1.64	1.62	1.60	1.57	1.55	1.53
500	1.74	1.71	1.69	1.66	1.64	1.62	1.61	1.59	1.56	1.54	1.52	1.50
1000	1.73	1.70	1.68	1.65	1.63	1.61	1.60	1.58	1.55	1.53	1.51	1.49
>1000	1.72	1.69	1.67	1.64	1.62	1.61	1.59	1.57	1.54	1.52	1.50	1.48

Table A.2 (*Continued*)

v_1/v_2	30	35	40	45	50	60	70	80	100	200	500	1000	>1000
3	8.62	8.60	8.59	8.59	8.58	8.57	8.57	8.56	8.55	8.54	8.53	8.53	8.54
4	5.75	5.73	5.72	5.71	5.70	5.69	5.68	5.67	5.66	5.65	5.64	5.63	5.63
5	4.50	4.48	4.46	4.45	4.44	4.43	4.42	4.42	4.41	4.39	4.37	4.37	4.36
6	3.81	3.79	3.77	3.76	3.75	3.74	3.73	3.72	3.71	3.69	3.68	3.67	3.67
7	3.38	3.36	3.34	3.33	3.32	3.30	3.29	3.29	3.27	3.25	3.24	3.23	3.23
8	3.08	3.06	3.04	3.03	3.02	3.01	2.99	2.99	2.97	2.95	2.94	2.93	2.93
9	2.86	2.84	2.83	2.81	2.80	2.79	2.78	2.77	2.76	2.73	2.72	2.71	2.71
10	2.70	2.68	2.66	2.65	2.64	2.62	2.61	2.60	2.59	2.56	2.55	2.54	2.54
11	2.57	2.55	2.53	2.52	2.51	2.49	2.48	2.47	2.46	2.43	2.42	2.41	2.41
12	2.47	2.44	2.43	2.41	2.40	2.38	2.37	2.36	2.35	2.32	2.31	2.30	2.30
13	2.38	2.36	2.34	2.33	2.31	2.30	2.28	2.27	2.26	2.23	2.22	2.21	2.21
14	2.31	2.28	2.27	2.25	2.24	2.22	2.21	2.20	2.19	2.16	2.14	2.14	2.13
15	2.25	2.22	2.20	2.19	2.18	2.16	2.15	2.14	2.12	2.10	2.08	2.07	2.07
16	2.19	2.17	2.15	2.14	2.12	2.11	2.09	2.08	2.07	2.04	2.02	2.02	2.01
17	2.15	2.12	2.10	2.09	2.08	2.06	2.05	2.03	2.02	1.99	1.97	1.97	1.96
18	2.11	2.08	2.06	2.05	2.04	2.02	2.00	1.99	1.98	1.95	1.93	1.92	1.92
19	2.07	2.05	2.03	2.01	2.00	1.98	1.97	1.96	1.94	1.91	1.89	1.88	1.88
20	2.04	2.01	1.99	1.98	1.97	1.95	1.93	1.92	1.91	1.88	1.86	1.85	1.84
22	1.98	1.96	1.94	1.92	1.91	1.89	1.88	1.86	1.85	1.82	1.80	1.79	1.78
24	1.94	1.91	1.89	1.88	1.86	1.84	1.83	1.82	1.80	1.77	1.75	1.74	1.73
26	1.90	1.87	1.85	1.84	1.82	1.80	1.79	1.78	1.76	1.73	1.71	1.70	1.69
28	1.87	1.84	1.82	1.80	1.79	1.77	1.75	1.74	1.73	1.69	1.67	1.66	1.66
30	1.84	1.81	1.79	1.77	1.76	1.74	1.72	1.71	1.70	1.66	1.64	1.63	1.62
35	1.79	1.76	1.74	1.72	1.70	1.68	1.66	1.65	1.63	1.60	1.57	1.57	1.56
40	1.74	1.72	1.69	1.67	1.66	1.64	1.62	1.61	1.59	1.55	1.53	1.52	1.51
45	1.71	1.68	1.66	1.64	1.63	1.60	1.59	1.57	1.55	1.51	1.49	1.48	1.47
50	1.69	1.66	1.63	1.61	1.60	1.58	1.56	1.54	1.52	1.48	1.46	1.45	1.44
60	1.65	1.62	1.59	1.57	1.56	1.53	1.52	1.50	1.48	1.44	1.41	1.40	1.39
70	1.62	1.59	1.57	1.55	1.53	1.50	1.49	1.47	1.45	1.40	1.37	1.36	1.35
80	1.60	1.57	1.54	1.52	1.51	1.48	1.46	1.45	1.43	1.38	1.35	1.34	1.33
100	1.57	1.54	1.52	1.49	1.48	1.45	1.43	1.41	1.39	1.34	1.31	1.30	1.28
200	1.52	1.48	1.46	1.43	1.41	1.39	1.36	1.35	1.32	1.26	1.22	1.21	1.19
500	1.48	1.45	1.42	1.40	1.38	1.35	1.32	1.30	1.28	1.21	1.16	1.14	1.12
1000	1.47	1.43	1.41	1.38	1.36	1.33	1.31	1.29	1.26	1.19	1.13	1.11	1.08
>1000	1.46	1.42	1.40	1.37	1.35	1.32	1.30	1.28	1.25	1.17	1.11	1.08	1.03

Table A.3   *F*-Table: Critical Values $\alpha = 0.01$

v_1/v_2	1	2	3	4	5	6	7	8	9	10	11	12
3	34.12	30.82	29.46	28.71	28.24	27.91	27.67	27.49	27.35	27.23	27.13	27.05
4	21.20	18.00	16.69	15.98	15.52	15.21	14.98	14.80	14.66	14.55	14.45	14.37
5	16.26	13.27	12.06	11.39	10.97	10.67	10.46	10.29	10.16	10.05	9.96	9.89
6	13.75	10.92	9.78	9.15	8.75	8.47	8.26	8.10	7.98	7.87	7.79	7.72
7	12.25	9.55	8.45	7.85	7.46	7.19	6.99	6.84	6.72	6.62	6.54	6.47
8	11.26	8.65	7.59	7.01	6.63	6.37	6.18	6.03	5.91	5.81	5.73	5.67
9	10.56	8.02	6.99	6.42	6.06	5.80	5.61	5.47	5.35	5.26	5.18	5.11
10	10.04	7.56	6.55	5.99	5.64	5.39	5.20	5.06	4.94	4.85	4.77	4.71
11	9.65	7.21	6.22	5.67	5.32	5.07	4.89	4.74	4.63	4.54	4.46	4.40
12	9.33	6.93	5.95	5.41	5.06	4.82	4.64	4.50	4.39	4.30	4.22	4.16
13	9.07	6.70	5.74	5.21	4.86	4.62	4.44	4.30	4.19	4.10	4.02	3.96
14	8.86	6.51	5.56	5.04	4.70	4.46	4.28	4.14	4.03	3.94	3.86	3.80
15	8.68	6.36	5.42	4.89	4.56	4.32	4.14	4.00	3.89	3.80	3.73	3.67
16	8.53	6.23	5.29	4.77	4.44	4.20	4.03	3.89	3.78	3.69	3.62	3.55
17	8.40	6.11	5.19	4.67	4.34	4.10	3.93	3.79	3.68	3.59	3.52	3.46
18	8.29	6.01	5.09	4.58	4.25	4.01	3.84	3.71	3.60	3.51	3.43	3.37
19	8.19	5.93	5.01	4.50	4.17	3.94	3.77	3.63	3.52	3.43	3.36	3.30
20	8.10	5.85	4.94	4.43	4.10	3.87	3.70	3.56	3.46	3.37	3.29	3.23
22	7.95	5.72	4.82	4.31	3.99	3.76	3.59	3.45	3.35	3.26	3.18	3.12
24	7.82	5.61	4.72	4.22	3.90	3.67	3.50	3.36	3.26	3.17	3.09	3.03
26	7.72	5.53	4.64	4.14	3.82	3.59	3.42	3.29	3.18	3.09	3.02	2.96
28	7.64	5.45	4.57	4.07	3.75	3.53	3.36	3.23	3.12	3.03	2.96	2.90
30	7.56	5.39	4.51	4.02	3.70	3.47	3.30	3.17	3.07	2.98	2.91	2.84
35	7.42	5.27	4.40	3.91	3.59	3.37	3.20	3.07	2.96	2.88	2.80	2.74
40	7.31	5.18	4.31	3.83	3.51	3.29	3.12	2.99	2.89	2.80	2.73	2.66
45	7.23	5.11	4.25	3.77	3.45	3.23	3.07	2.94	2.83	2.74	2.67	2.61
50	7.17	5.06	4.20	3.72	3.41	3.19	3.02	2.89	2.79	2.70	2.63	2.56
60	7.08	4.98	4.13	3.65	3.34	3.12	2.95	2.82	2.72	2.63	2.56	2.50
70	7.01	4.92	4.07	3.60	3.29	3.07	2.91	2.78	2.67	2.59	2.51	2.45
80	6.96	4.88	4.04	3.56	3.26	3.04	2.87	2.74	2.64	2.55	2.48	2.42
100	6.90	4.82	3.98	3.51	3.21	2.99	2.82	2.69	2.59	2.50	2.43	2.37
200	6.76	4.71	3.88	3.41	3.11	2.89	2.73	2.60	2.50	2.41	2.34	2.27
500	6.69	4.65	3.82	3.36	3.05	2.84	2.68	2.55	2.44	2.36	2.28	2.22
1000	6.66	4.63	3.80	3.34	3.04	2.82	2.66	2.53	2.43	2.34	2.27	2.20
>1000	1.04	4.61	3.78	3.32	3.02	2.80	2.64	2.51	2.41	2.32	2.25	2.19

Table A.3 (*Continued*)

v_1/v_2	13	14	15	16	17	18	19	20	22	24	26	28
3	26.98	26.92	26.87	26.83	26.79	26.75	26.72	26.69	26.64	26.60	26.56	26.53
4	14.31	14.25	14.20	14.15	14.11	14.08	14.05	14.02	13.97	13.93	13.89	13.86
5	9.82	9.77	9.72	9.68	9.64	9.61	9.58	9.55	9.51	9.47	9.43	9.40
6	7.66	7.61	7.56	7.52	7.48	7.45	7.42	7.40	7.35	7.31	7.28	7.25
7	6.41	6.36	6.31	6.28	6.24	6.21	6.18	6.16	6.11	6.07	6.04	6.02
8	5.61	5.56	5.52	5.48	5.44	5.41	5.38	5.36	5.32	5.28	5.25	5.22
9	5.05	5.01	4.96	4.92	4.89	4.86	4.83	4.81	4.77	4.73	4.70	4.67
10	4.65	4.60	4.56	4.52	4.49	4.46	4.43	4.41	4.36	4.33	4.30	4.27
11	4.34	4.29	4.25	4.21	4.18	4.15	4.12	4.10	4.06	4.02	3.99	3.96
12	4.10	4.05	4.01	3.97	3.94	3.91	3.88	3.86	3.82	3.78	3.75	3.72
13	3.91	3.86	3.82	3.78	3.75	3.72	3.69	3.66	3.62	3.59	3.56	3.53
14	3.75	3.70	3.66	3.62	3.59	3.56	3.53	3.51	3.46	3.43	3.40	3.37
15	3.61	3.56	3.52	3.49	3.45	3.42	3.40	3.37	3.33	3.29	3.26	3.24
16	3.50	3.45	3.41	3.37	3.34	3.31	3.28	3.26	3.22	3.18	3.15	3.12
17	3.40	3.35	3.31	3.27	3.24	3.21	3.19	3.16	3.12	3.08	3.05	3.03
18	3.32	3.27	3.23	3.19	3.16	3.13	3.10	3.08	3.03	3.00	2.97	2.94
19	3.24	3.19	3.15	3.12	3.08	3.05	3.03	3.00	2.96	2.92	2.89	2.87
20	3.18	3.13	3.09	3.05	3.02	2.99	2.96	2.94	2.90	2.86	2.83	2.80
22	3.07	3.02	2.98	2.94	2.91	2.88	2.85	2.83	2.78	2.75	2.72	2.69
24	2.98	2.93	2.89	2.85	2.82	2.79	2.76	2.74	2.70	2.66	2.63	2.60
26	2.90	2.86	2.82	2.78	2.75	2.72	2.69	2.66	2.62	2.58	2.55	2.53
28	2.84	2.79	2.75	2.72	2.68	2.65	2.63	2.60	2.56	2.52	2.49	2.46
30	2.79	2.74	2.70	2.66	2.63	2.60	2.57	2.55	2.51	2.47	2.44	2.41
35	2.69	2.64	2.60	2.56	2.53	2.50	2.47	2.44	2.40	2.36	2.33	2.31
40	2.61	2.56	2.52	2.48	2.45	2.42	2.39	2.37	2.33	2.29	2.26	2.23
45	2.55	2.51	2.46	2.43	2.39	2.36	2.34	2.31	2.27	2.23	2.20	2.17
50	2.51	2.46	2.42	2.38	2.35	2.32	2.29	2.27	2.22	2.18	2.15	2.12
60	2.44	2.39	2.35	2.31	2.28	2.25	2.22	2.20	2.15	2.12	2.08	2.05
70	2.40	2.35	2.31	2.27	2.23	2.20	2.18	2.15	2.11	2.07	2.03	2.01
80	2.36	2.31	2.27	2.23	2.20	2.17	2.14	2.12	2.07	2.03	2.00	1.97
100	2.31	2.27	2.22	2.19	2.15	2.12	2.09	2.07	2.02	1.98	1.95	1.92
200	2.22	2.17	2.13	2.09	2.06	2.03	2.00	1.97	1.93	1.89	1.85	1.82
500	2.17	2.12	2.07	2.04	2.00	1.97	1.94	1.92	1.87	1.83	1.79	1.76
1000	2.15	2.10	2.06	2.02	1.98	1.95	1.92	1.90	1.85	1.81	1.77	1.74
>1000	2.13	2.08	2.04	2.00	1.97	1.94	1.91	1.88	1.83	1.79	1.76	1.73

(*continued*)

Table A.3 (*Continued*)

v_1/v_2	30	35	40	45	50	60	70	80	100	200	500	1000	>1000
3	26.50	26.45	26.41	26.38	26.35	26.32	26.29	26.27	26.24	26.18	26.15	26.13	26.15
4	13.84	13.79	13.75	13.71	13.69	13.65	13.63	13.61	13.58	13.52	13.49	13.47	13.47
5	9.38	9.33	9.29	9.26	9.24	9.20	9.18	9.16	9.13	9.08	9.04	9.03	9.02
6	7.23	7.18	7.14	7.11	7.09	7.06	7.03	7.01	6.99	6.93	6.90	6.89	6.89
7	5.99	5.94	5.91	5.88	5.86	5.82	5.80	5.78	5.75	5.70	5.67	5.66	5.65
8	5.20	5.15	5.12	5.09	5.07	5.03	5.01	4.99	4.96	4.91	4.88	4.87	4.86
9	4.65	4.60	4.57	4.54	4.52	4.48	4.46	4.44	4.42	4.36	4.33	4.32	4.32
10	4.25	4.20	4.17	4.14	4.12	4.08	4.06	4.04	4.01	3.96	3.93	3.92	3.91
11	3.94	3.89	3.86	3.83	3.81	3.78	3.75	3.73	3.71	3.66	3.62	3.61	3.60
12	3.70	3.65	3.62	3.59	3.57	3.54	3.51	3.49	3.47	3.41	3.38	3.37	3.36
13	3.51	3.46	3.43	3.40	3.38	3.34	3.32	3.30	3.27	3.22	3.19	3.18	3.17
14	3.35	3.30	3.27	3.24	3.22	3.18	3.16	3.14	3.11	3.06	3.03	3.01	3.01
15	3.21	3.17	3.13	3.10	3.08	3.05	3.02	3.00	2.98	2.92	2.89	2.88	2.87
16	3.10	3.05	3.02	2.99	2.97	2.93	2.91	2.89	2.86	2.81	2.78	2.76	2.75
17	3.00	2.96	2.92	2.89	2.87	2.83	2.81	2.79	2.76	2.71	2.68	2.66	2.65
18	2.92	2.87	2.84	2.81	2.78	2.75	2.72	2.71	2.68	2.62	2.59	2.58	2.57
19	2.84	2.80	2.76	2.73	2.71	2.67	2.65	2.63	2.60	2.55	2.51	2.50	2.49
20	2.78	2.73	2.69	2.67	2.64	2.61	2.58	2.56	2.54	2.48	2.44	2.43	2.42
22	2.67	2.62	2.58	2.55	2.53	2.50	2.47	2.45	2.42	2.36	2.33	2.32	2.31
24	2.58	2.53	2.49	2.46	2.44	2.40	2.38	2.36	2.33	2.27	2.24	2.22	2.21
26	2.50	2.45	2.42	2.39	2.36	2.33	2.30	2.28	2.25	2.19	2.16	2.14	2.13
28	2.44	2.39	2.35	2.32	2.30	2.26	2.24	2.22	2.19	2.13	2.09	2.08	2.07
30	2.39	2.34	2.30	2.27	2.25	2.21	2.18	2.16	2.13	2.07	2.03	2.02	2.01
35	2.28	2.23	2.19	2.16	2.14	2.10	2.07	2.05	2.02	1.96	1.92	1.90	1.89
40	2.20	2.15	2.11	2.08	2.06	2.02	1.99	1.97	1.94	1.87	1.83	1.82	1.81
45	2.14	2.09	2.05	2.02	2.00	1.96	1.93	1.91	1.88	1.81	1.77	1.75	1.74
50	2.10	2.05	2.01	1.97	1.95	1.91	1.88	1.86	1.82	1.76	1.71	1.70	1.69
60	2.03	1.98	1.94	1.90	1.88	1.84	1.81	1.78	1.75	1.68	1.63	1.62	1.60
70	1.98	1.93	1.89	1.85	1.83	1.78	1.75	1.73	1.70	1.62	1.57	1.56	1.54
80	1.94	1.89	1.85	1.82	1.79	1.75	1.71	1.69	1.65	1.58	1.53	1.51	1.50
100	1.89	1.84	1.80	1.76	1.74	1.69	1.66	1.63	1.60	1.52	1.47	1.45	1.43
200	1.79	1.74	1.69	1.66	1.63	1.58	1.55	1.52	1.48	1.39	1.33	1.30	1.28
500	1.74	1.68	1.63	1.60	1.57	1.52	1.48	1.45	1.41	1.31	1.23	1.20	1.17
1000	1.72	1.66	1.61	1.58	1.54	1.50	1.46	1.43	1.38	1.28	1.19	1.16	1.12
>1000	1.70	1.64	1.59	1.56	1.53	1.48	1.44	1.41	1.36	1.25	1.16	1.11	1.05

Table A.4 Critical Values of Studentized Range Distribution (q) for Familywise ALPHA = 0.05

Denominator DF	Number of Groups (Treatments)							
	3	4	5	6	7	8	9	10
1	26.98	32.82	37.08	40.41	43.12	45.40	47.36	49.07
2	8.33	9.80	10.88	11.73	12.43	13.03	13.54	13.99
3	5.91	6.83	7.50	8.04	8.48	8.85	9.18	9.46
4	5.04	5.76	6.29	6.71	7.05	7.35	7.60	7.83
5	4.60	5.22	5.67	6.03	6.33	6.58	6.80	7.00
6	4.34	4.90	5.31	5.63	5.90	6.12	6.32	6.49
7	4.17	4.68	0.06	5.36	5.61	5.82	6.00	6.16
8	4.04	4.53	4.89	5.17	5.40	5.60	5.77	5.92
9	3.95	4.42	4.76	5.02	5.24	5.43	5.60	5.74
10	3.88	4.33	4.65	4.91	5.12	5.30	5.46	5.60
11	3.82	4.26	4.57	4.82	5.03	5.20	5.35	5.49
12	3.77	4.20	4.51	4.75	4.95	5.12	5.26	5.40
13	3.73	4.15	4.45	4.69	4.88	5.05	5.19	5.32
14	3.70	4.11	4.41	4.64	4.83	4.99	5.13	5.25
15	3.67	4.08	4.37	4.60	4.78	4.94	5.08	5.20
16	3.65	4.05	4.33	4.56	4.74	4.90	5.03	5.15
17	3.63	4.02	4.30	4.52	4.71	4.86	4.99	5.11
18	3.61	4.00	4.28	4.49	4.67	4.82	4.96	5.07
19	3.59	3.98	4.25	4.47	4.65	4.79	4.92	5.04
20	3.58	3.96	4.23	4.45	4.62	4.77	4.90	5.01
21	3.57	3.94	4.21	4.42	4.60	4.74	4.87	4.98
22	3.55	3.93	4.20	4.41	4.58	4.72	4.85	4.96
23	3.54	3.91	4.18	4.39	4.56	4.70	4.83	4.94
24	3.53	3.90	4.17	4.37	4.54	4.68	4.81	4.92
25	3.52	3.89	4.15	4.36	4.53	4.67	4.79	4.90
26	3.51	3.88	4.14	4.35	4.51	4.65	4.77	4.88
27	3.51	3.87	4.13	4.33	4.50	4.64	4.76	4.86
28	3.50	3.86	4.12	4.32	4.49	4.63	4.75	4.85
29	3.49	3.85	4.11	4.31	4.48	4.61	4.73	4.84
30	3.49	3.85	4.10	4.30	4.46	4.60	4.72	4.82
31	3.48	3.84	4.09	4.29	4.45	4.59	4.71	4.81
32	3.48	3.83	4.09	4.28	4.45	4.58	4.70	4.80
33	3.47	3.83	4.08	4.28	4.44	4.57	4.69	4.79
34	3.47	3.82	4.07	4.27	4.43	4.56	4.68	4.78
35	3.46	3.81	4.07	4.26	4.42	4.56	4.67	4.77
36	3.46	3.81	4.06	4.26	4.41	4.55	4.66	4.76
37	3.45	3.80	4.05	4.25	4.41	4.54	4.66	4.76
38	3.45	3.80	4.05	4.24	4.40	4.53	4.65	4.75
39	3.45	3.80	4.04	4.24	4.39	4.53	4.64	4.74
40	3.44	3.79	4.04	4.23	4.39	4.52	4.63	4.74
41	3.44	3.79	4.04	4.23	4.38	4.52	4.63	4.73
42	3.44	3.78	4.03	4.22	4.38	4.51	4.62	4.72
43	3.43	3.78	4.03	4.22	4.37	4.50	4.62	4.72

Table A.4 (*Continued*)

Denominator DF	Number of Groups (Treatments)							
	3	4	5	6	7	8	9	10
44	3.43	3.78	4.02	4.21	4.37	4.50	4.61	4.71
45	3.43	3.77	4.02	4.21	4.36	4.49	4.61	4.71
46	3.43	3.77	4.02	4.21	4.36	4.49	4.60	4.70
47	3.42	3.77	4.01	4.20	4.36	4.49	4.60	4.70
48	3.42	3.76	4.01	4.20	4.35	4.48	4.59	4.69
49	3.42	3.76	4.01	4.19	4.35	4.48	4.59	4.69
50	3.42	3.76	4.00	4.19	4.34	4.47	4.58	4.68
51	3.41	3.76	4.00	4.19	4.34	4.47	4.58	4.68
52	3.41	3.75	4.00	4.18	4.34	4.47	4.58	4.67
53	3.41	3.75	3.99	4.18	4.33	4.46	4.57	4.67
54	3.41	3.75	3.99	4.18	4.33	4.46	4.57	4.67
55	3.41	3.75	3.99	4.18	4.33	4.46	4.57	4.66
56	3.41	3.75	3.99	4.17	4.33	4.45	4.56	4.66
57	3.40	3.74	3.98	4.17	4.32	4.45	4.56	4.66
58	3.40	3.74	3.98	4.17	4.32	4.45	4.56	4.65
59	3.40	3.74	3.98	4.17	4.32	4.44	4.55	4.65
60	3.40	3.74	3.98	4.16	4.31	4.44	4.55	4.65
61	3.40	3.74	3.98	4.16	4.31	4.44	4.55	4.64
62	3.40	3.73	3.97	4.16	4.31	4.44	4.55	4.64
63	3.40	3.73	3.97	4.16	4.31	4.43	4.54	4.64
64	3.39	3.73	3.97	4.16	4.31	4.43	4.54	4.64
65	3.39	3.73	3.97	4.15	4.30	4.43	4.54	4.63
66	3.39	3.73	3.97	4.15	4.30	4.43	4.54	4.63
67	3.39	3.73	3.97	4.15	4.30	4.43	4.53	4.63
68	3.39	3.73	3.96	4.15	4.30	4.42	4.53	4.63
69	3.39	3.72	3.96	4.15	4.30	4.42	4.53	4.62
70	3.39	3.72	3.96	4.14	4.29	4.42	4.53	4.62
71	3.39	3.72	3.96	4.14	4.29	4.42	4.53	4.62
72	3.38	3.72	3.96	4.14	4.29	4.42	4.52	4.62
73	3.38	3.72	3.96	4.14	4.29	4.41	4.52	4.62
74	3.38	3.72	3.95	4.14	4.29	4.41	4.52	4.61
75	3.38	3.72	3.95	4.14	4.29	4.41	4.52	4.61
76	3.38	3.72	3.95	4.14	4.28	4.41	4.52	4.61
77	3.38	3.71	3.95	4.13	4.28	4.41	4.51	4.61
78	3.38	3.71	3.95	4.13	4.28	4.41	4.51	4.61
79	3.38	3.71	3.95	4.13	4.28	4.40	4.51	4.60
80	3.38	3.71	3.95	4.13	4.28	4.40	4.51	4.60
81	3.38	3.71	3.95	4.13	4.28	4.40	4.51	4.60
82	3.38	3.71	3.95	4.13	4.28	4.40	4.51	4.60
83	3.38	3.71	3.94	4.13	4.27	4.40	4.50	4.60
84	3.37	3.71	3.94	4.13	4.27	4.40	4.50	4.60
85	3.37	3.71	3.94	4.12	4.27	4.40	4.50	4.60
86	3.37	3.71	3.94	4.12	4.27	4.39	4.50	4.59
87	3.37	3.70	3.94	4.12	4.27	4.39	4.50	4.59

Table A.4 (*Continued*)

Denominator DF	Number of Groups (Treatments)							
	3	4	5	6	7	8	9	10
88	3.37	3.70	3.94	4.12	4.27	4.39	4.50	4.59
89	3.37	3.70	3.94	4.12	4.27	4.39	4.50	4.59
90	3.37	3.70	3.94	4.12	4.27	4.39	4.50	4.59
91	3.37	3.70	3.94	4.12	4.26	4.39	4.49	4.59
92	3.37	3.70	3.94	4.12	4.26	4.39	4.49	4.59
93	3.37	3.70	3.93	4.12	4.26	4.39	4.49	4.59
94	3.37	3.70	3.93	4.11	4.26	4.38	4.49	4.58
95	3.37	3.70	3.93	4.11	4.26	4.38	4.49	4.58
96	3.37	3.70	3.93	4.11	4.26	4.38	4.49	4.58
97	3.37	3.70	3.93	4.11	4.26	4.38	4.49	4.58
98	3.37	3.70	3.93	4.11	4.26	4.38	4.49	4.58
99	3.37	3.70	3.93	4.11	4.26	4.38	4.49	4.58
100	3.37	3.70	3.93	4.11	4.26	4.38	4.48	4.58

INDEX

Repeated Measures Design for Empirical Researchers, First Edition. J. P. Verma.
© 2016 John Wiley & Sons, Inc. Published 2016 by John Wiley & Sons, Inc.